Communicating Mathematics in the Digital Era

Communicating Mathematics in the Digital Era

Edited by
J.M. Borwein
E.M. Rocha
J.F. Rodrigues

A K Peters, Ltd.
Wellesley, Massachusetts

Editorial, Sales, and Customer Service Office

A K Peters, Ltd.
888 Worcester Street, Suite 230
Wellesley, MA 02482
www.akpeters.com

Library of Congress Cataloging-in-Publication Data

Communicating mathematics in the digital era / edited by J.M. Borwein, E.M. Rocha, J.F. Rodrigues.
p. cm.
Includes bibliographical references and index.
ISBN 978-1-56881-410-0 (alk. paper)
1. Mathematics–Data processing–Congresses. 2. Libraries and electronic publishing–Congresses. 3. Image processing–Digital techniques–Mathematics –Congresses. I. Borwein, Jonathan M. II. Rocha, E. M. (Eugenio M.) III. Rodrigues, José-Francisco.

QA76.95.C59 2008
621.36′7–dc22

2008022183

Printed in India
12 11 10 09 08 10 9 8 7 6 5 4 3 2 1

To our children,
both young and adult,
and to one very special grandson.

Contents

Preface

The Digital Era has dramatically changed the ways that researchers search, produce, publish, and disseminate their scientific work. This is certainly true for mathematicians. These processes are still rapidly evolving due to improvements in Information Science (e.g., information architecture, archiving, long-term preservation), new achievements in Computer Science Technologies (e.g., XML based formats, data mining, visualization, web tools) and initiatives such as DML and open access journals (e.g., DOAJ), digitization projects (e.g., EMANI, GDZ, JSTOR, NUMDAM), scientific reference catalogs (e.g., MathSciNet, ZMath, Google Scholar), and digital repositories (e.g., EPRINTS, DSpace). These changes have prompted many mathematicians to play an active part in the developments of the Digital Era, and have led mathematicians to promote and discuss new ideas with colleagues from other fields, such as technology developers and publishers, so as to enhance the paradigms and mechanisms for producing, searching, and exploiting scientific and technical scholarship in mathematics.

While not a traditional proceedings, this book reflects many of the contributions regarding the subjects described above that were delivered and discussed at the ICM 2006 satellite meeting entitled "Communicating Mathematics in the Digital Era" (CMDE2006), which took place at the University of Aveiro in Portugal, August 15–18, 2006. The main speakers were J. Borwein (CEIC-IMU, Chair), T. Bouche (NUMDAM, Mini-DML Project, Gallica-Math, CEDRAM), B. Cipra (JPBM 2005 Communication Awardee), J. Ewing (American Mathematical Society, Executive Director), P. Ion (W3C HTML-Math Working Group, Co-chair), J. Kiernan (IBM Almaden Research Center), B. Wegner (ELibM/EMIS,

Chair), and E. Weisstein (Wolfram Inc., MathWorld Creator). The meeting had participants from 15 countries: Canada, China, Czech Republic, Finland, France, Germany, Italy, Japan, Lithuania, Norway, Poland, Portugal, Spain, the United Kingdom, and the United States of America.

We have organized the contributions to the volume as follows. After an overview on disseminating mathematical knowledge, Part I comprises eight articles on Electronic Publishing and Digital Libraries. Part II contains six articles on the technology of dissemination, while Part III offers five varied snapshots on related educational and cultural issues. To all contributors we offer our sincere thanks.

The editors wish to express their deep appreciation to Chris Hamilton and especially to Mason Macklem who were responsible for much of the technical editing of this book.

Finally, for their financial support of the CMDE2006 meeting, we most gratefully acknowledge the following research groups: Centro de Estudos em Optimização e Controlo (CEOC) and Matemática e Aplicações (UIMA) of the Department of Mathematics of the University of Aveiro, and the Fundação para a Ciência e Tecnologia (FCT), co-financed by FEDER. We wish to thank also the Sociedade Portuguesa de Matemática (SPM), the Real Sociedad Matemática Española (RSME), the European Mathematical Society (EMS) and the Committee on Electronic Information and Communication (CEIC) of the International Mathematical Union (IMU).

Jonathan M. Borwein
Eugénio M. Rocha
José Francisco Rodrigues

1 Electronic Publishing and Digital Libraries

Disseminating and Preserving Mathematical Knowledge

E.M. Rocha and J.F. Rodrigues

At the dawn of the digital age, the importance of the documentation of mathematical knowledge assumes a strategic and decisive character, both in current research as well as in the transmission to future generations. Of course, this is something that happens in all branches of knowledge. Documentary resources such as databases, publication archives, or electronic resources (digital books or simple web pages), as well as increasingly powerful search engines, are essential instruments not only for science but also for the economy and the information society. However, for mathematics and all who use it, the importance of documentation is strengthened by the generality and permanence intrinsic to the subject, to which is due its unique situation amid the other branches of human knowledge.

Historically, the transmission of mathematics is indissociably connected to the technology specific to each period, traversing the last twenty-five centuries of civilization. From manuscripts to the printing press, from classic libraries to digital repositories, from verbal or postal transmission to electronic communication, the technological revolution we are living through presents mathematicians, and all professionals who use mathematics with new challenges and new opportunities. Mathematical documents are currently created and transmitted in digital form, thus allowing inclusion of other elements besides the traditional text and static images. Computers and the Internet have radically altered our ways of communicating and sharing ideas and results, enhancing the potential of the traditional forms of human thought and intelligence.

There are many players in this complex universe: authors, referees, editors, publishers, libraries, repositories, World Wide Web servers, readers, financing foundations and agencies, universities, institutes, and scientific societies. At the turn of the century, an ambitious idea that gave rise to the Worldwide Digital Library of Mathematics (WDML) [url:57] started to mobilize efforts and to generate projects of national and international collaboration for the purpose of creating a gigantic virtual library to be used by the global mathematical and scientific community. However, the set of changes and adaptations that constitutes the revolution in information technology has not yet completely settled down the new ways of sharing and developing the mathematical sciences. Mathematics is particularly dependent on its literature, and this might easily be jeopardized by the increasing ease of electronic publication and communication. So, mathematicians and mathematical societies must be aware of current progress and be able to contribute to the exploration of the new technologies, in order to preserve and to transmit mathematics.

Electronic Publications in Mathematics

Five centuries separate the first typographical edition of Euclid's *Elements*, printed in Venice in 1482 by Erhard Ratdolt, from the announcement in 1978 of the TEX language by its creator, the mathematician and computer scientist Donald Knuth [160]. This should be contrasted with the five years that elapsed between this event and the adoption by the *Transactions and the Proceedings of the American Mathematical Society* (AMS) of this digital system of mathematical composition. Although it was neither the first nor the only electronic typesetting system, TEX became stable and universally accepted by the mathematical community, despite the existence of certain dialects like LATEX and AMSTEX. For mathematicians today it constitutes almost a reinvention of the alphabet and has become an essential instrument of communication.

Nowadays, practically all submissions of mathematical works, be they articles or monographs, are using TEX and its diverse forms of visualization (e.g., PostScript, PDF), which not only immensely facilitate their publication but also cut publication costs substantially, "democratizing" mathematical writing and typography. Moreover, the new versions of TEX permit the inclusion of hyperlinks in the electronic text, which allow new searches associated to hypertext and electronic publication in cyberspace.

While the 1980s marked the beginning of the systematic use of TEX in mathematical typography, the following decade saw the widespread use of electronic mail and, above all, the invention of the World Wide Web. These new electronic resources have brought new methods of communication such as repositories of pre-publications and electronic journals. Scientific journals are the privileged vehicles for the publication of mathematical research, being an established form that goes back two centuries. The *Journal für die Reine und Angewandte Mathematik* was established in 1826 in Berlin by A.L. Crelle, and the *Journal de Mathématiques Pures et Appliquées* in 1836 by J. Liouville in Paris, and both are still published today, being periodicals of highest scientific reputation. The journal tradition is based on a publishing system of critical pre-reading (refereeing), decision (by the editorial board), and scientific publication (by the publisher). This enriches the value of each author's work in a way that is difficult to quantify; the mathematical community values this system and considers it irreplaceable, while looking for new forms of maintaining this system in the digital age.

The first appearance of solely electronic mathematics journals took place as far back as 1992 with the ephemeral *Ulam Quarterly* that ceased publication with the third volume four years later, and with the *Electronic Journal of Differential Equations* and the *Electronic Transactions on Numerical Analysis* that have been published regularly since 1993. Today, there are over ten titles directly accessible from the EMIS/ELibM [url:106] catalog (Figure 1) of the European Mathematical Society (EMS). However, if the acceptance of this new type of journal took some time, with the gradual digitization of all mathematics journals and their availability (at least for subscribers) on the Internet, the distinction starts to disappear. Beyond the advantage of accessibility, the solely electronic journals show that the cost of mathematical publications can be minimized. It must be said that a great number of journals, as for example, the ones edited by the recently started EMS-Publishing House, already include hyperlinks in their electronic editions, for formulas and for bibliographies, including references to MathSciNet [url:69] or Zentralblatt MATH [url:111].

On the other hand, the last decade has witnessed an enormous reorganization of the publishing industry, with acquisitions and mergers, and simultaneously, a constant and paradoxical increase of subscription prices for those journals belonging to the main scientific publishing companies, which between them control about 60% of the main journals. For example, D. Knuth [161], in a significant October 2003 letter

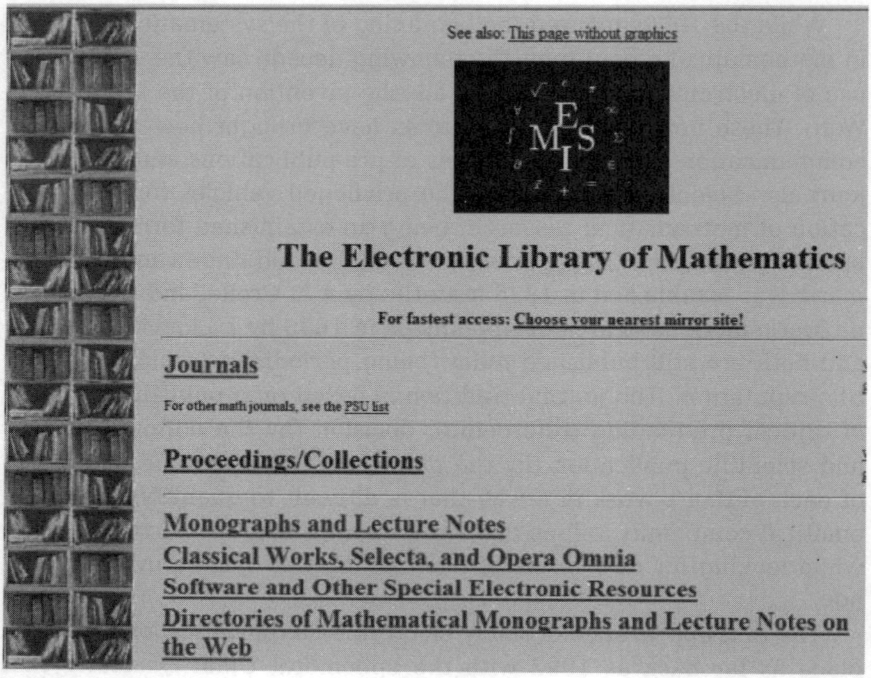

Figure 1. Screen capture of the homepage of the Electronic Library of Mathematics.

to the publishing committee of the *Journal of Algorithms* (that he had helped to found in 1979 for Academic Press), noted with indignation typical of the scientific community that after many years of a steady 25 to 30 cents cost per page, the price had risen to 50 cents per page in 2002 after the acquisition of the journal by Elsevier.

In another interesting reaction to the evolution of subscription costs for mathematics journals belonging to commercial publishing companies, the mathematician Joan Birman [37], of Columbia University, New York, observed that in the face of the increasing costs for these journals, the mathematicians/publishers of the main mathematics journals can and must react, looking for other publishing companies that offer the same quality of publication for more reasonable prices. Birman gave as an example the collective resignation of 50 members of the editorial committee of the *Journal of Logic Programming*, after drawn out and fruitless negotiations on its price with Elsevier, followed by the foundation of the new journal *Theory and Practice of Logic Programming*, published by Cambridge University Press with a 55% price reduction. Another example is the collective resignation of nine editors of *Topology*, a well-established journal since 1962 and published by Elsevier since 1994. In their letter of August 2006, they state: "We believe that the price, in combination with Elsevier's policies for pricing mathematical journals more generally, has had a significant and damaging effect on *Topology*'s reputation in the mathematical research community, and that this is likely to become increasingly serious and difficult, indeed impossible, to reverse in the future." [url:82] The London Mathematical Society has recently released the following notice: "It is with great pleasure that we announce the launch of a new journal, to be called the *Journal of Topology*."

In her article of the year 2000, Birman classified the existing mathematical research journals into four categories:

1. university journals, for example, the *Annals of Mathematics* (University of Princeton and the Institute for Advanced Studies) or the *Annali della Scuola Normale Superiore di Pisa* (Italy), that accept exchanges;
2. journals belonging to scientific societies, such as *Transactions of the American Mathematical Society* or *Portugaliæ Mathematica*, of the *Sociedade Portuguesa de Matemática* (SPM);
3. journals belonging to university presses, for example, the *European Journal of Applied Mathematics* that is published by Cambridge University Press—almost indistinguishable from type 1, except that they do not accept exchanges;
4. journals belonging to commercial publishing companies.

Birman estimated that among the best journals, percentages within each of these categories would be 17%, 13%, 10% and 60%, respectively.

More precise data has been supplied by John Ewing [103], Executive and Publishing Director of the AMS, in a cautious article that nevertheless supports experimentation with new technologies to find better ways to communicate mathematics. Ewing relates that, in 2001, Mathematical Reviews indexed or analyzed 51,721 articles from 1,172 distinct journals. Of these journals, 591 (about 50%) were "cover to cover" (journals considered to be only about mathematics) containing 30,924 articles (60% articles) of all mathematics, while the remaining 581 were classified as "other" (i.e., multidisciplinary or out of the mathematical "mainstream") contained the remaining 40% of mathematics articles. However, that year, only 46 (4%) of the journals, containing only 1,272 articles (2.5%), were considered "primarily-electronic," even though approximately two thirds of all the articles, i.e., about

34,000, possessed hyperlinks, meaning that at least they were available in electronic version. Also of interest is the information Ewing relays about the raised costs of subscriptions to commercial journals that, owned by great publishing companies, continue to generate high profits and to dominate the market. In 1991, only 24% of the journals were commercial and published 38% of the articles, but in 2001, some 30% (349) of the journals were commercial and published 48% (25,008) of the mathematics articles! Ewing thus concluded that alternative electronic editions in mathematics, not only the electronic journals but also the pre-publication repositories like arXiv [url:5] and MPRESS [url:39], still represent under 10% of all recent mathematical publication and do not present immediate danger to the great publishing companies, which have adapted to electronic publication by offering new services with the subscriptions.

However, it is not yet clear how the significant fact that Grigory Perelman (who was awarded a Fields Medal at the 2006 ICM Madrid Congress for "his contributions to geometry and his revolutionary insights into the analytical and geometric structure of the Ricci flow") has chosen to publish his three fundamental papers that solved the Poincaré conjecture only as preprints in arXiv will affect the current situation in the publication of mathematical research. These papers, as well as other commentary and references, can be accessed from a Clay Mathematics Institute page [url:89].

On the other hand, Ewing [103] relates an extraordinarily interesting statistic about one of the new characteristics of the Mathematical Reviews, which consists of referring to the list of citations in articles of certain journals, which included, in 2002, about 340,000 citations made since 1998. Of these, 53% referred to articles published before 1990, and more than 28% were articles previous to 1980, which is particularly significant given the increase of mathematics articles. This fact highlights the singularity of the importance and permanence of mathematical literature for the advance of this science, in contrast with the other sciences, including physics, whose literature has a more ephemeral character.

In Portugal, for instance, a country that is not central in the history of mathematics, the mathematical press is five centuries old, going back to 1496 with the publication, in Leiria, of Abraham Zacutus' astronomical tables—the Almanach Perpetuum—and with the publication of the Tratado da Pratica d' Arismetyca, by Gaspar Nicolás, in Lisbon in 1516. In spite of its reduced dimension and impact on the recent development of mathematics, the first periodic Portuguese mathematical journal dates from 1877 when the *Journal das Ciencias Mathematicas e Astronomicas* was founded in Coimbra by F. Gomes Teixeira. This journal was published for a quarter of a century. Nowadays, only one mathematics research journal is published in Portugal, namely *Portugaliæ Mathematica*, which was founded in 1937 and is currently published by the SPM. *Portugaliæ Mathematica* has used computer typesetting in TEX since Volume 47 in 1987, has had an electronic version on the Internet since 2001 in the EMIS/ELibM [218], and has seen its first 50 volumes retrodigitized by the National Library of Portugal in a successful experiment conducted by J.R. Borbinha.

Despite the relative sluggishness in the process of substitution of paper by digital, and their continuing coexistence in the final version, paper has almost disappeared in the initial phases of submission, refereeing, and production of scientific articles, where the use of email and TEX-based files is almost universal. This confirms that electronic publication is a well-established and consolidated fact in mathematics communication in the twenty-first century. Such a situation was already predictable in the 1980s: one of the authors of this article remembers how in 1988/1989 he edited, in Lisbon, a collective book for Birkhäuser, for which more than half of the twenty contributions arrived electronically in TEX.

If the practices of electronic production and distribution of mathematical literature are now accepted by the scientific community, both the long-term preservation of archival documents and the viability of access costs constitute serious problems, so far without viable solutions, and thus are a worry for both librarians and scientists.

The Publication Model

In fact, the current publication process is in crisis and is undergoing profound transformation. Its economic models are in question and there is controversy between the traditional reader-payer model and the author-payer model; this latter is on the rise as it is the foundation of *Open Access* (OA).

The main idea of OA is to break down the trade barriers to literature, on the one hand allowing free access to information and, on the other, minimizing the costs of dissemination by delegating the responsibility to the authors or to public or private institutions. The declarations of Budapest [url:163] and Berlin [url:196] are well known, and as can be seen in the Directory of Open Access Journals [url:98] more than 1800 journals currently exist only in OA, of which more than

60 are in the area of mathematics. Guidelines for the sustainability of journals in the OA regime [url:164, url:156] can be found on the Internet.

One of the strongest arguments for OA is based on the observation that research is financed by governmental institutions and, therefore, society should have the right of access to the scientific content paid for by their taxes. Another argument points out that public money should be used for the common good, not for the benefit of private publishing companies that acquire the copyright to research and profit from it, without paying the authors. This idea underpins the *Declaration on Access to Research Data from Public Funding*, an agreement signed under the auspices of the Organization for Economic Co-operation and Development (OECD).

In January 2006 the European Commission published the "Study on the Economic and Technical Evolution of the Scientific Publication Markets of Europe" [url:16], the first recommendation of which is to "guarantee public access to publicly-funded research results shortly after publication." The study resulted from analysis of the current scholarly journal publication market, as well as consultation with all the major stakeholders within the scholarly communication process (e.g., researchers, funders, publishers, librarians, research policy-makers). Moreover, the final report on scientific publication of the European Research Advisory Board recommends that the commission consider mandating that all researchers funded under FP7 deposit their publications resulting from EC-funded research in an open access repository as soon as possible after publication, to be made openly accessible within six months at the latest [url:15].

This model is, however, far from finding consensus not only among enterprise specialists, like J.J. Esposito [105], who anticipates a general increase in the cost of publications for research in the OA model, but also among members of the mathematical community. John Ewing [104], aware that 75% of the annual income of the AMS comes from its publications, considers that the solution is not in free access to publications, but in creating a demand to lower the prices of journal subscriptions for librarians and academics. J. Ball and J. Borwein [19] affirm that the OA model is not well-adapted to mathematics, for it might place its publications at the mercy of the administrators of the universities and other institutions, as well as at the mercy of other, richer, disciplines. This may cause discrimination among mathematicians according to whether they can fund their publications in cheaper or more expensive journals.

The Digital Library of Mathematics

Mathematical literature has a long tradition of organization and archiving, in part due to its accumulative nature, in part to the necessity felt by mathematicians to create databases and reviewing journals as Zentralblatt MATH (associated with the EMS) and the *Mathematical Reviews* (of the AMS), currently accessible on the Internet. This tradition goes back to the nineteenth century, from which the *Jahrbuch über die Fortschritte der Mathematik* (JFM) remains an outstanding example, having reviewed over 200,000 mathematical publications in the 68 volumes published between 1868 and 1942. This journal is currently the subject of a project (Electronic Research Archive for Mathematics) of online digitization, having been distinguished recently with a 2005 award from the Special Libraries Association/Physics-Astronomy-Mathematics Division. In May 2005, the JFM site [url:108] already had made use of 17,035 external hyperlinks to facsimiles of original documents to be found in various libraries and repositories in Germany, France, and the United States. With the gradual retrodigitization of mathematical documents and the placing online of global digital mathematical document repositories, all mathematical literature will one day hopefully be at the distance of a click for any person, anywhere, at any time.

It is clear that this vision, likely to become reality more rapidly than might be imagined, still has many obstacles in front of it, including the technical problems of accessibility and long-term archive preservation. Among the current initiatives, it is worth mentioning EMANI (the Electronic Mathematics Archives Network Initiative) [url:105], a project that the German geometer Bernd Wegner, one of its main coordinators, describes in [268].

The current rate of mathematical publication is enormous; the Zentralblatt MATH databases (Figure 2), published since 1931, contain more than two million mathematical articles, growing at a rate of about 80,000 articles per year and covering about 2,300 journals, serial collections, conference proceedings, collected works, and books. These numbers show why the traditional libraries struggle with problems of space, financing, and other serious hindrances.

The European Mathematical Society, in cooperation with the Facinformationszentrum Karlsruhe, opened the EMIS (European Mathematical Information Service) portal, which has been available on the Internet since June 1995 [147]. With free access and with dozens of "mirrors" in the whole world, this portal constitutes a most useful ser-

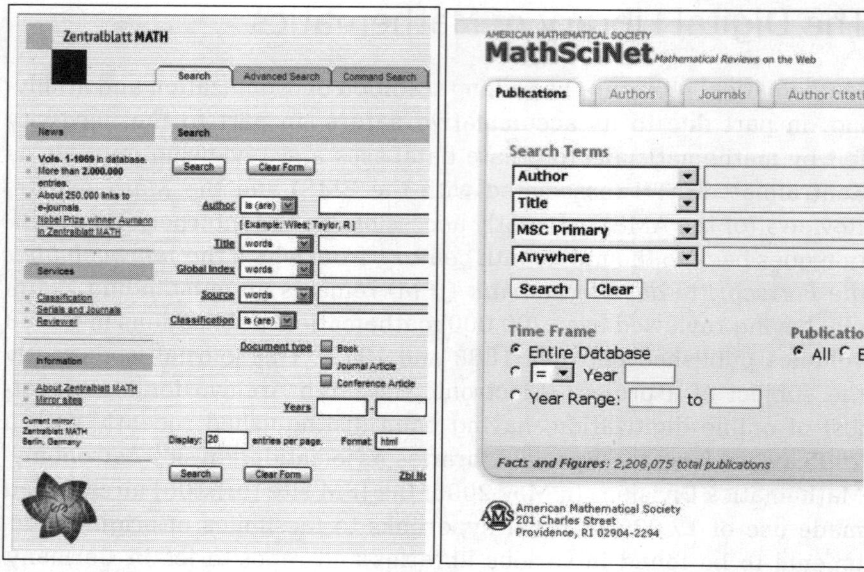

Figure 2. Screen captures of the Zentralblatt MATH database (left), and the American Mathematical Society's MathSciNet database (right).

vice to the international community with three main components: an electronic library, a collection of databases (some available through subscription), and projects. In 1995, the EMS made available in the EMIS an Electronic Mathematics Library (ELibM) [url:106], supervised by its Committee for Electronic Publication. From the 12 mathematics journals that were available in November 1995, it progressed to the current catalog of 82 journals (March 2007), as well as a long list of proceedings/collections (including the *Proceedings of the International Congress of Mathematicians*, 1998 (Berlin)), monographs and lecture notes, classic works (including the collected works of Hamilton and Riemann), electronic and computational resources and, recently, a free access directory of books and monographs on the Internet. All the material may be accessed for free, except for journals that are within the so-called "moving wall," i.e., a period of restricted access—usually of three to five years—after which the articles become available for free. This period allows the publishing companies to recoup investments made in the publication process. In the follow-up to the International Congress of 1998 in Berlin, the International Mathematical Union (IMU) created a Committee for Electronic Information and Communication (CEIC). In the context of the initiatives planned for the World Mathematical Year, in 2000, mathematicians had started to speak of the necessity of digitizing the historical mathematical literature in order to make it available and to relate it with the current mathematical production, which tends to be all digital, in one way or another. Philippe Tondeur, professor at the University of Illinois and director of the division of Mathematics in the National Science Foundation in the US (NSF) between 1999 and 2002, supported the idea of establishing an international project for massive digitization of articles with the aim of creating the Digital Mathematics Library (DML). To this end, the NSF financed the Library of the University of Cornell for a year, supporting a discussion group on the subject, currently known as the DML Planning Group. To encourage the project, John H. Ewing (executive director of the AMS) published an article [102] in 2002 in which he considers the possibility, using current technology, of digitizing all the old mathematical documents, estimated by Keith Dennis of Cornell to be about 50 million pages, and making them available on the Internet. One of the key points made by Ewing that was received positively by the scientific community was the observation that mathematics is, in fact, the ideal branch for learning to carry out an enterprise like the DML. It should be pointed out that the average life of a mathematics article is approximately ten years, whereas in other sciences it could be as little as six months [url:120].

The great aim of the DML is, therefore, to bring together all mathematical literature and to make it available through a centralized portal, accessible via the Internet, that coordinates the various sources. To reach this goal, it will be necessary to digitize all paper-based literature (a gigantic task that will only be possible if it is decentralized and performed with common or compatible standards), to create a system that integrates all literature in digital format, and to establish connections with databases such as those of MathReviews (AMS) and Zentralblatt (EMS) (Figure 2); in short, to adopt technical norms that adequately guarantee the correct dissemination, permanent update and long-term preservation of materials and, of course, the integrity of copyright.

As recognized by the IMU, the DML "is a vital effort for the mathematical community," and presently there exists a worldwide movement of coordination, but also of competition, in order to execute the multiple projects necessary to this collective enterprise. This process is necessarily complex and is distributed among several partners. It involves the production of digital documents, and relies on the establishment of standards for their technical specifications, and standards for the metadata, interoperability and interconnection, the archive, and update and maintenance. The problems of cost and of economic models of sustainability, the question of intellectual property and copyright, and the international management and coordination of digitization, must also be addressed.

(Retro)Digitization

Retrodigitization is the name of the process that consists of creating digital copies of documents that exist only in paper format. One of the aims of the Digital Mathematics Library (DML) is to retrodigitize all the existing mathematical legacy that is not yet in digital format. The retrodigitization process is basically composed of two phases: obtaining a digital version of the paper document and structuring the acquired information in a useful and accessible database. In the first phase, an image is usually created for each page of the document and kept in PDF or DjVu format. After that, the metadata are created and text recognition software (OCR) is used to construct indices that allow for electronic searching of the content. The process is not expensive, considering that many projects of massive retrodigitization have sent their documents to be digitized in countries where wages are very low, so the majority of costs correspond to the introduction of bibliographical data and equipment. The estimated cost per page is on average 2 euros, 10% of which is the cost of the actual digitization.

In recent years, the idea of the DML has been realized through various projects and initiatives, so that the number of retrodigitized journals available online is already significant. In the DML maintained by Ulf Rehmann [url:138] there are links to 2,247 digitized books (> 514,565 pages) and 222 journals (> 4,018,752 pages); in some cases, complete access is paid for by the user. This work is the result of regional projects financed by national agencies or mathematics societies from countries such as Germany (EMANI, GDZ), France (GALLICA, NUMDAM), United States (JSTOR), Poland (BWM), and Portugal (PM-SPM in collaboration with the National Library), and include the following:

- Biblioteka Wirtualna Matematyki (BWM) [url:40] with three journals (1888–1993), including *Fundamenta Mathematicae* (1920–1993) and *Studia Mathematica* (1929–1964);
- GALLICA [url:25] with 230 mathematics entries, including *Liouville's Journal* (1836–1932), the four volumes of the *Histoire de Mathématiques* by J.F. Montucla (1799–1802), and several works

by Huygens, Euler, Fourier, Cauchy, Darboux, and Jordan, among others;

- Göttinger Digitalisierungszentrum (GDZ-Göttingen) [url:168] with 28 journals and several monographs (1777–1997), including the complete works of Gauss, Klein, and Hilbert, the editions of 1898 and 1939 of the *Encyklopädie der Mathematischen Wissenschaften mit Einschlussiher Anwendunge*, and the *Zentralblatt für Mathematik und ihre Grenzgebiete* (1931–1978);
- Journal STORage (JSTOR) [url:122] with 28 journals (1800–2002), including the *Annals of Mathematics* (1884–1997), the *Journal of the AMS* (1988–1997), the *Journal of the Royal Statistical Society* (1988–1998), the *Mathematische Annalen* (1869–1996), *SIAM Review* (1959–1997) and the *Proceedings of the AMS* (1950–1997);
- NUMérisation de Documents Anciens Mathématiques (NUMDAM) [url:147] with six journals (1864–2000), including the complete text of the work *Eléments de Géometrie Algébrique* by A. Grothendieck and J. Dieudonné, the journals *Annales de l'Institut Fourier* (1949–1997) and *Annales Scientifiques de l'École Normale Supérieure* (1864–1997);
- PM-SPM (Portuguese project) [url:50], containing the retrodigitization of the *Portugaliæ Mathematica* (1937–1993).

More examples of recent and ongoing projects, like the RusDML for the Russian mathematics literature, the DML-CZ in the Czech Republic, and the DML-E digitization project in Spain, are described in this book.

Despite the relevance of the amount of information available, the current implementation of the DML does not yet satisfy the premise of centralized access. The various projects operate in a relatively isolated and independent way, with the data being made available in very distinct forms. As observed by Ewing in [102], it would make sense to create a coordinating entity to define guidelines and technical practices to follow for all such projects. However, the projects that already exist continue to work without input from, or rule-setting by, any higher authority. Consequently, there are visible disparities between services and functionalities available to the user of each project. For example, access to GDZ is free but it does not perform OCR on the images, and so the search for keywords in the content of its documents is impossible. JSTOR, which allows an integrated search and makes available various functionalities such as obtaining documents in TIFF, PDF, and PostScript formats, is a commercial venture, charging thousands of dollars for access. NUMDAM, which contains links to the MathSciNet and Zentralblatt MATH, is a free service based in the University of Grenoble and supported by the French CNRS, but, like the remaining projects, it was in the beginning limited to national journals. In the technical context, despite the absence of a definite and globally accepted norm, several entities publish recommendations on good practice to adopt in the digitization process, for example Minerva eEurope [url:142] and the Committee of Electronic Information and Comunications of the International Mathematics Union (CEIC/IMU) [url:83]. However, mathematics presents concrete difficulties resulting from the dispersion and diversity of documents. As mathematical journals were (and are) published variously by commercial publishing houses, university publishing companies, scientific societies, departments of mathematics, and even by groups of mathematicians, it thus becomes difficult to acquire and transport the original copies and to negotiate copyrights. Currently, digitization and the subsequent creation of digital libraries are so important for the preservation of the cultural and scientific heritage that in a document dated April 28, 2005 addressed to the Presidency of the European Council, six Heads of State and Government defended the creation of a European virtual library. The European Commission congratulated itself for this plan and will contribute to its accomplishment through the emblematic initiative called i2010 Digital Libraries [url:20].

Search and Metadata

State of the art scientific activity depends heavily on the search and consultation of bibliographical material. However, the vast amount, and the continual growth, of scientific literature constitute a difficulty for authors and researchers in identifying and using all the information relevant to their work. According to the article [168], the most-cited articles are those of easiest access (for example, those that are online). In particular, in the field of computer science, in a sample of 119,924 conference articles, the average number of offline articles cited is 2.74, while the average number of online articles cited is 7.03. Note that, in this scientific area, conference articles carry more weight than journal articles, having an average acceptance of 10%. In mathematics, no similar studies have been published, but they would probably follow this standard. Forms of global and integrated search are ever more important and necessary for the development and progress of scientific research.

Mathematical Databases

In addition to historical catalogs, such as the *Jahrbuch über die Fortschritte der Mathematik* (JFM) (1868–1944) and the *Répertoire Bibliographique des Sciences Mathématiques* (1894–1912) [url:36], there exists the current Mathematics Education Database [url:107], accessible through the EMIS, which constitutes an online bibliographical database of mathematical education that is associated with the German journal ZDM (*Zentralblatt für Didaktik DES Mathematik*) and contains over 119,344 citations (1976–June, 2007).

But the main support tools for current mathematical research are the databases of MathSciNet and Zentralblatt MATH (Figure 2), each one with around two million entries and each with its own search engine in the respective online versions. These not only allow document searches, but also supply some statistics on publications.

For example, to help the passage of Portugal from Group I to Group II of the IMU, the Portuguese National Commission of Mathematics conducted a quantitative survey of publications referenced in MathSciNet (books, journals, and proceedings), counting per year those publications with at least one Portuguese institution-based author in the period from 1985 to 2004 (search "Inst Code=p- *" on October 26, 2005).

Google Scholar and Google Print

Not long ago, Google made available the services "Google Scholar" (GScholar) [url:53] and "Google Print" (GPrint) [url:47], still in an experimental phase. GScholar is a bibliographic search mechanism, counting, currently, more than 500 million references to scientific documents. Among other interesting characteristics, it allows for a search that includes articles, summaries and citations; locates a document in a library (when not available online); and allows the grouping of several versions of a work (for example, preprint, journal article). GPrint, in turn, permits the search and visualization of book pages with the search keywords appearing in color, brings the user to a shopping portal on the web, or verifies the book's existence in libraries belonging to the WorldCat [url:149]. The number of pages accessible for visualization depends on the status of the book: in free access, the whole book is obtainable; conditional access permits visualization of two pages (either side); unavailable access offers only bibliographical references. A sample document available in GPrint is shown in Figure 3. Google encourages publishers of all disciplines to make available all references

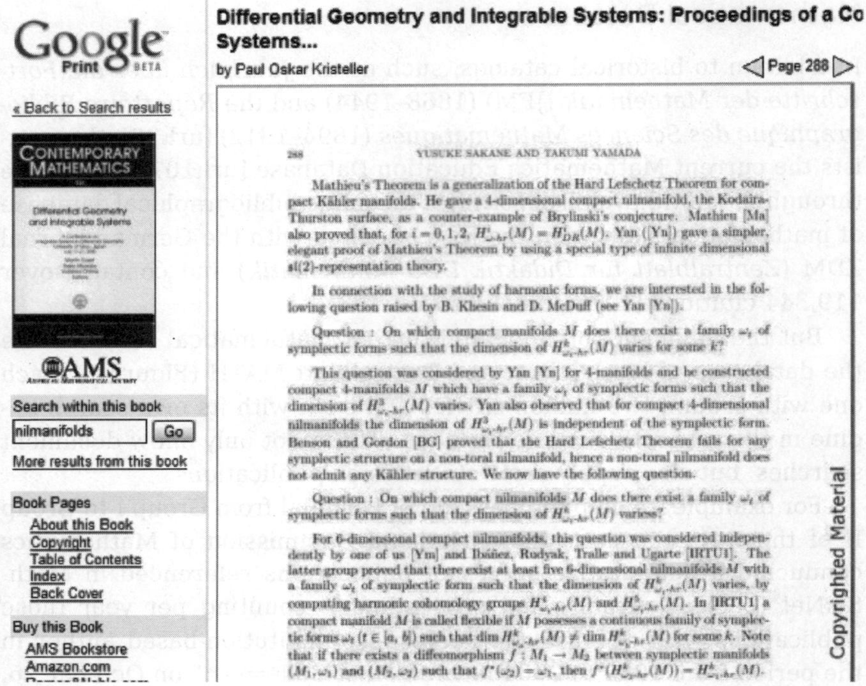

Figure 3. Screenshot of a sample document available through GPrint.

to their material [url:54]. Total control is delegated to the publishers concerning publication policies, but they must make available at least a summary for each article. GScholar promises to deal with all violations of copyright that are communicated to it, as defined by the Digital Millennium Copyright Act [url:92].

Google also encourages libraries to use GScholar [url:55]. This service allows members of a university to find, through GScholar, a link to the catalog entry of a document in its library, whenever this is available. The system also connects to WorldCat, permitting, for example, a search for the closest library where the document is available. In GScholar it is possible to personalize some parameters, for example, to configure a default localization; allowing a user outside the university campus to accomplish searches as though on campus. The authentication of users and control of access to documents are the entire responsibility of the libraries.

These services are a sample of what will be possible to access on the Internet when the millions of books from five great libraries (of the

Universities of Oxford, Harvard, Stanford, and Michigan; and of the New York Public Library), being digitized by Google, become available in its database.

National Digital Libraries–The "b-on" Example

National Digital Libraries (NDLs) are usually created by the corresponding National Science Foundations to provide organized access to high quality resources and tools, building bridges among the scientific, research, and educational communities; and between private sector and public interests, e.g., by providing access to resources such as publishers' journal articles. An example of this is the National Science Digital Library established by the National Science Foundation (NSF) [url:43], in 2000, as an online library that directs users to exemplary resources for science, technology, engineering, and mathematics. Another example is the Portuguese NDL called "Online Library of Knowledge" (b-on) [url:72], which allows full access to the main sources of international scientific knowledge for all members of the national academic community, research and development centers, as well as other public institutions. Moreover, b-on allows an allocation of costs through a global negotiation with the scientific publishing companies.

A year and a half after its creation, b-on, with reference to more than sixteen thousand publications and four million visualizations of integral texts, has altered its portal to improve its usability, and has increased the available functionalities. These include interfaces adapted to the user's profile (beginner/regular/expert), parametrization of alert, and context help, among others. B-on is a federated search engine (based on Metalib), used by the universities of Harvard and Stanford, and by several Finnish universities. There are currently 69 institutions using b-on, up from forty-eight institutional users in b-on's early days. Content is distributed by areas: art and humanities (11.58%), science and technology (20.69%), health sciences (17.99%), social sciences (38.52%), and physics, chemistry and mathematics (11.22%); this is according to data supplied by the Agency for the Society of Knowledge (UMIC). The costs of its functioning have been estimated at 7.5 million Euros in 2004 and 10.8 million Euros in 2005, 56.6% of this being paid by the State and the remainder by the institutional users.

The importance of b-on is recognized internationally. Besides the Economist Intelligence Unit having cited it as a model in a report on cases of success in the use of structural funds for the countries of the enlargement, this initiative has been presented in several international forums. The interim evaluation report for the Lisbon Strategy [url:104] by the High Level Group, presided over by Wim Kok, recognizes the importance of the Society of Knowledge in the design to transform Europe into the most competitive space in the world.

However, despite the enormous progress that recourse to b-on represents for mathematicians in Portugal, important gaps still exist in the area of mathematics, such as access to JSTOR, the journals of the EMS and the AMS and, in particular, the MathReviews. On the other hand, b-on's impact on the classical libraries has yet to be evaluated, as well as the possible future consequences of the lack of paper versions of the main journals in case of insufficient financing, or discontinuation, of some online services.

Final Considerations

Considering the present community tendencies and choices, it is reasonable to expect that the way we conceive, develop, communicate, and disseminate mathematics will change in the near future. These changes will be brought forward by several initiatives, streams of knowledge, and groups of study. On the presentation level, it is not only the typesetting system LaTeX that is evolving. Other products are incorporating new ways of typing and displaying mathematics, using new structural (and semantic) markup formats, such as MathML (OpenMath), sTeX, OMDoc, and Microsoft Word 2007 (which has new features that distinguish it from previous versions). Merging these with symbolic computation systems, computer algebra systems (e.g., Mathematica, Maple, Wiris, and specific Java applets) are progressively changing the paradigm of being only isolated applications that help us with the development process, to being also material that we can incorporate in the scientific output of our research work. In fact, scientific content is still rather traditional from the interactive and presentation points of view. Although some electronic journals already allow authors to add to their publications certain types of multimedia content, the scientific published material in digital form is far smaller than the whole amount published. Nevertheless, the World Wide Web is already the largest single resource of mathematical knowledge, containing thousands of contributions collected by several projects, e.g., MathWorld, Wikipedia, and the new project PlanetMath. Other collaborative projects are emerging that incorporate not only new forms of collecting and displaying mathematics, but also new ways of organizing the knowledge. These are based on processes of codifying the semantics in a machine understandable form, thereby enabling us to build ontologies characterizing the concepts and relations involved. Hypothetically, these new technologies will allow us to extend the lines of collaboration; automatically check proofs (e.g., see openCyc and Mizar projects); and improve the classification, searching, and ranking mechanisms (e.g., enabling us to search by concept and not by keywords, such as "Is there a theorem/result that is adequate for this problem?" or "Which theorems/results are connected with this one?"). On an even more ambitious level, these new technologies will, in the long-term future, enable us to build a Worldwide Library of Mathematical Knowledge (WLMK). As the Worldwide Digital Mathematics Library (WDML) has the purpose of creating a virtual library of digital documents, the WLMK has the purpose of creating a virtual library of structured knowledge (independent of its supporting digital format), to be used by the global mathematical and scientific community.

The Digital Downside

J. Ewing

Scholars and librarians usually focus on the advantages of electronic journals—faster processing, reduced costs, and new features (such as searching and linking)—and there are indeed many such advantages. But like all technology, electronic journals have a downside as well. Most people have ignored this downside because these problems are presently little more than annoyances. But we should understand the downside in order to prevent these small annoyances from turning into big crises in the future. In one or two cases, it may already be too late.

Here are five problems with electronic journals—problems that arise when the miraculous new technology is combined with the human frailties of carelessness, greed, myopia, dogmatism, and infatuation.

1 Careless Scholars

In the digital age, we can do things we could never do before. Here are some examples from the recent literature of things we can do:

- A journal posts an article in January; in April, without any notice, the editors replace the article with a "corrected" version.
- A journal posts an article in July; in November, the publisher simply removes the article (without notifying the authors!).
- A journal posts an article in April, but the author posts a corrected and substantially changed version on a well-known server in October, with an indication that the article appeared in April, but no indication that it was different.

Mathematics relies on its literature for its underpinnings; its literature is interconnected. Imagine a world in which one percent of the mathematical literature is affected in the above way, in which one percent of the articles one finds are not the "authentic" versions. Over time, as work based on faulty references spreads, the fraction of unreliable literature will increase. Experts may be able to overcome this, but *nonexperts* will be overwhelmed. We ignore this potential crisis at the peril of future generations of mathematicians.

These sloppy practices occur because new technology allows us to do things never before possible. That doesn't mean we *should* do them! To prevent this from destroying the scholarly literature, we need to insist on high standards. There are two ways to handle this—back-linking or forward-linking—and both require discipline. Every author regrets publishing mistakes (as do publishers!), but we have to resist the temptation to hide them.

2 Big Deals

The electronic age has made it possible for *big* publishers to offer *big* deals. Here's the way they work. Rather than subscribing to journals one by one, an institution is offered electronic access to a huge package of journals across many fields (many of which were previously unavailable to the institution). Initially, the cost of this package is comparable to the cost of the publisher's journals to which the institution previously subscribed, and (the publisher points out) it's always *far* less than the total cost of the individual subscriptions. This seems to be a wonderful opportunity for the institution, and the vice-president for information (the one who negotiated the deal) crows about the fiscal prowess that brought about this arrangement.

Big deals are now offered by Elsevier (an innovator in this area), Springer, Wiley, Blackwell, and Taylor & Francis. In a recent survey of US research libraries, 93% indicated that they held bundles with at least one of these publishers. Just about half (49%) held bundles with at least four. Wiley, Elsevier, and Springer have achieved 70% market penetration.[1]

When asked about their motivation, most said that it was a "good return on investment" and that the "alternatives . . . were prohibitively expensive."

[1]The survey collected data from 89 of the 123 member libraries of the Association of Research Libraries during November and December of 2005.

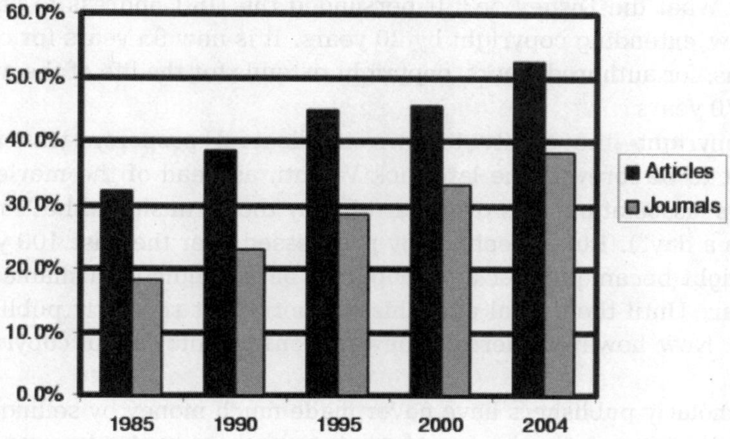

Figure 1. Percentage of commercial journals among all mathematics journals.

But these institutions have paid a heavy price for that "good investment." Such big deals are almost always multi-year contracts that do not allow cancellations or changes. The extra titles are often of marginal value to scholars. Most importantly, decisions about what is purchased are made at a high level, far removed from scholars themselves. In the end, big deals make it more difficult for scholars to make sensible decisions about journals based on price and need. Of course, big deals give the big publishers a substantial advantage over smaller publishers, which is their real purpose.

Big deals are hard to resist, although a few prominent libraries have done so. But the commercial publishers are winning. In 1985, roughly a third of mathematics articles were in commercial journals; by 2004, over half the articles were commercial.[2] We need to fight back.

3 Walt Disney

The Walt Disney story is a metaphor. The first Mickey Mouse cartoon was produced and copyrighted in 1928. By the late 1990s, the term of copyright for Mickey Mouse (then 75 years for corporate works) was about to expire in 2003. The Disney corporation was making lots of money on the character, partly because new technology allowed new

[2]These data were derived from *Mathematical Reviews*.

uses. What did Disney do? It persuaded the US Congress to change the law, extending copyright by 20 years. It is now 95 years for corporations; for authored works, copyright extends for the life of the author plus 70 years.

Copyright—the ability to own intellectual property—was never meant to be forever (the late Jack Valenti, as head of the movie producers' association, was overheard to say the term should be "forever minus a day"). But as technology progressed over the past 400 years, copyright became longer and longer . . . barely indistinguishable from forever. Until the digital age, this did not affect scholarly publishing much. Now, however, there are new reasons to worry about copyright's reach.

Scholarly publishers have never made much money by selling journal back volumes. In the age of print, publishers kept a few copies of old journals to sell to libraries when they wanted to replace missing volumes or (rarely) to start a new collection. Typically, such sales of back volumes amounted to one to three percent of journal revenues. Publishers expected to recover their initial costs (and make a profit) by selling *current* subscriptions, not by selling back volumes.

Now, however, like the Disney corporation, publishers see an opportunity to make money on their old material using new technology. They want to sell their journals twice, once as a current subscription and a second time as a collection of backfiles. There are many variations on this scheme, some merely sell the backfiles *with* the current subscription, pointing out that it makes the subscription more valuable (and hence more expensive). But the central point is that publishers carefully control access to the backfiles.

Of course, publishers who digitize their back volumes want to recover their costs. But rather than find a way to pay the *one-time* cost of digitization, and then make the material freely available, they want to continue making money next year, and the year after that, and after that, and on and on into the future. This includes publishers of every kind, not just the commercial ones.

This doesn't make sense for Mickey Mouse, but it makes even less sense for scholarship. By "owning" scholarship and restricting access indefinitely, publishers make it impossible to realize the dream of connecting the large body of the past scholarly literature to the present. This makes our present journals *less* valuable, not more, and ultimately hurts the publishers themselves.

Copyright was invented to make publishing profitable. But no one—no publisher, no author, no one—needs to hold copyright for more than two or three decades. We ought to make a pact that everything goes into the public domain after 28 years.[3]

4 Mindless Accountants

Using flawed statistics to make (equally flawed) decisions is not new. But it is so, so much easier to do in the digital age. Many people have written about the "impact factor," which is more and more frequently misused, but I want to talk about another troublesome statistic—journal usage, which is even more dangerous than the impact factor.

Librarians all over the world are insisting on usage statistics for journals, which roughly translates into the number of downloads of various articles over a given period of time. They claim this is necessary in order to measure the value of journals and to make decisions about subscriptions. Publishers in general seem happy to oblige—almost eager in many cases.

This is a hopelessly naive and dangerous game. What is the meaning of such usage statistics? What does it mean that some article has been downloaded 100 times? Have people read it 100 times? Surely not—you don't *read* every item on which you click while browsing the web, so why would scholars read every article they download? And which is more valuable, an article downloaded once a week for ten years, or one downloaded 520 times in its first month? What about caching and the many flaws in browser software that give rise to faulty counts? There are many questions, but almost no answers—just the demand for usage statistics in order to measure value.

Making decisions using ignorant accounting has always been a bad idea, but in this case it may have some disastrous consequences. If librarians are really going to measure value by clicks, surely publishers will force users to click more often. It would be foolish to give away abstracts or references or even bibliographic data, for example, if this leads to fewer clicks and hence less value. And if value equals clicks, then why on earth would publishers let authors post copies of papers anywhere *except* on the publisher's website. The more liberal a publisher's policy in this regard, the more the publisher risks losing value. Making decisions by using flawed usage statistics will inevitably shift publishers' practices, all in the wrong direction. We need to explain

[3]The term of 28 years was standard for more than two centuries after copyright was invented, and therefore has historical meaning. See *Notices of the AMS* 51(3), March 2004, p. 309.

this to those who dogmatically claim that value can be measured by a few flawed numbers.

5 Fads and Fashions

A corrupted literature, a literature controlled by a handful of giant publishers, a literature hidden away forever, a literature shaped by nutty accountants using flawed statistics—all these are potential crises caused by the advent of electronic publishing. None is insurmountable, but each is worrisome. And so with these worries pressing upon us, what do we advance as our most pressing issue in the new electronic age? Access—open access to the literature.

On the face of it, this is bizarre. In the digital age, scholars have more access to the literature than ever before. When an institution subscribes to a journal, the articles are now delivered straight to a user's desktop. Finding articles is far easier than ever before using any number of search tools. Articles can be downloaded, printed, and (dare we admit?) sent to others via email. Even when a user's institution does not subscribe to a journal, the user can see the abstract and (often) the list of references. This allows scholars to decide whether the article is useful and then to send email to the author to ask for a copy (or simply to find it elsewhere on the web—recent changes in publisher policies make this easy). And many publishers provide access to older articles without any subscription at all. None of this was possible in the print-only world—none of it! In the print world, good access to the scholarly literature was restricted to a few institutions, and near-universal access was unheard of. This has all changed.

Nonetheless, instead of focusing on the four potential crises mentioned above, the scholarly community has decided to focus on access, the one aspect of the scholarly literature that has already been greatly improved in the digital age.

I have spent a lot of time trying to understand why people seem obsessed with access, and I have come to realize that there is no simple answer. In part, it is human nature: when things improve, we want to improve them still more. Having a taste of increased access, people want completely unfettered access. In part, this is because the call to "open access" is simple to understand. Scholars are not willing to invest time in wrestling with the tougher issues of scholarly publishing mentioned above, which are messy and sometimes hard to unravel. But there is a more subtle, and more insidious, reason for the obsession with access: whenever technology changes the world around us, we become more susceptible to fads and fashions. New technology opens up new opportunities, and new opportunities bring forth opportunists—zealous people who promote their own special causes.

On the face of it, increased access is surely not a bad thing: it is hard to argue against having more access to scholarship. On the other hand, it *can* be bad if it causes us to ignore the real problems we face, and it can be tragic if new enticing technology combines with an irresistible fad to mislead us into acting against our own interests.

Open access has had both effects on mathematics. When planning for our digital future, we spend most of our time talking about access (already greatly improved), and almost no time talking about the integrity of scholarship, copyright issues, foolish bureaucrats who use faulty statistics, or (worst of all!) avaricious publishers who have created a crisis in scholarly publishing. Instead, we talk about access. And, of course, those avaricious publishers are delighted by the distraction.

We also formulate ideas that are clearly bad for mathematics. The author-pay model for journals simply does not work for mathematicians. Our funding levels do not support it and they never will, at least at the levels of medical sciences. We will always be at a competitive disadvantage to other scientists in an author-pay world.

The self-archiving model of open access is clearly bad for mathematics, which more than any other discipline depends on the long-term survival of a reliable web of scholarship—into the distant future many decades from now. And preprint servers, without any obvious source of long-term funding, are not much more attractive.

The government-funded model of open access is also clearly bad for mathematics, which has never competed well with the other sciences for government largess. Besides, the budget for the government-funded model seems to assume a stable publishing environment for the long term, without the need to invest in ever changing technology. Surely this is short-sighted. We will have to invest even more in the coming decade than we have in the past in order to keep up with changing technology. Will the government really do that investing?

Indeed, every model of open access that has been proposed is clearly detrimental to the broad interests of mathematicians. But open access has become an obsession for many that has blinded otherwise thoughtful people into acting against their own interests. That is a real downside to new technology that may do the most damage in the long term. This is a downside that already affects us.

Don't believe me? Just check out those large commercial publishers who are all beginning to embrace open access. They are creating separate corporate units, just to promote and implement open access. They can do that—they have the resources. They know that no matter how the business model changes, they will be able to take advantage of it to make money—lots of it. Surely publishers who have raised subscription prices for years will feel equally free to raise author charges in the future. Change is good for large corporations, who have the resources to invest in change; it is a lot harder for the little guys who operate on a tight budget and make a tiny profit. This is especially bad for mathematics, which has traditionally relied on more small and independent journals than any other discipline. If open access is so great for us—if open access is about to solve all our problems—how come the commercial publishers are jumping on the bandwagon?

6 Conclusions

Are there disadvantages to electronic journals? Of course there are. Should we advocate abandoning the digital age because of those disadvantages? Of course not. And in any case, the digital age is here to stay, no matter what we advocate. But as we move forward into the new age, we need to make clear the principles that should underlie our scholarly literature.

I believe in formulating clear policies about the integrity of scholarship. I believe in fighting to keep decision making in the hands of scholars. I believe that no one should own ideas for too long—scholarship belongs to all of us and moving walls should be universal. And I believe that scholars, and not accountants, should measure the value of scholarship.

But I believe most of all that we must stay focused on these *real* problems as the world changes around us, and not be distracted by fads—especially fads that will hurt mathematics (and scholarship more generally) in the long run.

The digital age will dramatically change the way in which we disseminate scholarship in the future ... but we have to guide that change so that it makes things better and not worse.

Implementing Electronic Access for an Independent Journal

Technical Issues, Business Decisions, Legal Matters

K. Kaiser

We discuss from the perspective of a typical independent journal the transition from "print" to "print plus electronic." Besides technical issues concerning implementation of restricted access, decisions on pricing electronic editions and legal matters concerning the license had to be made. The *Houston Journal of Mathematics* wanted to find a middle ground between its obligation to disseminate knowledge as freely as possible and its need to generate enough revenue in order to maintain its independence. As a journal that has been around for about 30 years, the creation of an electronic archive was a natural additional step. I will explain how we integrated access to backfiles in our subscription model. Finally, I will try to justify my positive outlook for independent journals in general.

1 Introduction

The shift from "print" to "print plus electronic" has created for all scholarly journals a host of new and challenging problems. This is especially true when a journal opts for restricted electronic access. Then not only technical problems concerning implementation have to be resolved, but also issues concerning marketing and pricing have to be addressed. The purchase of copyrighted electronic material requires in most cases a license agreement. Because current laws concerning (fair) use of copyrighted material do not apply to electronic files, it is up to the publisher to decide to what extent electronic files can be used. And in case of license violations, in absence of criminal abuse, the publisher has to decide what kind of legal actions, if any, should be taken.

Commercial journals employ specialists for electronic matters, together with people who know about web design and advertising; and there are business executives and lawyers. In contrast, independent journals are run by only a few people, often only by one academic person, and a very small staff for secretarial and accounting work. It would seem that independent journals are hopelessly disadvantaged in this electronic age. However, in my experience as the Managing Editor of a typical independent research journal, the *Houston Journal of Mathematics* (HJM), I can say that this is just not true and the purpose of my article is to support this claim.

First, I am going to describe in simple terms how restricted access can be implemented. Interestingly enough, the technical choices we made have actually influenced the terms and policies of our license.

Currently, copyright and usage issues for documents in electronic format are in a legal state of limbo, with librarians and publishers often on opposite sides. Some librarians feel that they can extrapolate certain copyright laws to electronic issues, despite the fact that these laws were crafted long before the price of copying became a nonissue; and publishers have been accused of trying to sell the work of their authors indefinitely, without carrying any further expenses. The HJM license tries to find a middle ground between the interests of subscribers, and our need to protect subscription revenue.

Commercial journals stress electronic subscriptions. For a variety of reasons, a few major US libraries have already decided to subscribe only to the electronic editions of the more expensive Elsevier and Springer journals. This trend may lead to the eventual disappearance of print editions. How will this affect the small academic publishers? Some people predict that future technical developments will be too demanding for smaller publishers and in order to survive, they must band together, or join larger organizations. I have some thoughts on what kind of future cooperation with other publishers or agencies HJM might consider as beneficial and would support.

2 Starting a Website: First Decisions

HJM advanced during the year of 1998 from a "print" to a "print plus electronic" journal. HJM had already established its web presence the year before with freely accessible tables of contents, abstracts, and a comprehensive index. Thus, in order to complete the process, we only had to provide links from titles to document files. Like most other mathematics journals, we chose the PDF file format for the web, as well as for printing.

From a business perspective, print and electronic editions share the very same upfront expenses for production of the final PDF files. For print editions one only has to add the costs of printing and mailing, for electronic editions one must add the nontrivial work of updating the website.

It should be obvious that in order to create and maintain a full-fledged website, a journal needs to have access to a computer network that can accommodate the special needs of electronic journal publishing. For HJM, this was no problem because of excellent computing facilities and support within the mathematics department. Thus, there was no need for HJM to find an electronic journal publisher.

However, having the technical means to run a website is, for a journal, only the first requirement for acting as its own publisher. The site also needs regular updates. We wanted to have for HJM a complete and very informative site, but also one that was easy to maintain. For this reason, our site is mostly text-based. We also tried to minimize the number of internal links. Of course, titles need to be linked to files, but we put abstracts and titles all on the same page. For every issue, this saves us the work of creating about 20 additional webpages, together with their links.

Right from the beginning, we asked authors to provide us with auxiliary abstracts that are suitable for web posting. That is, web abstracts have to be in plain English, self-contained, and with all mathematics in UNICODE (with no $\TeX$ code whatsoever).

Before issues are sent to the printer, all authors receive the final PDF file for last-minute corrections and updating the references. On our website, issues that have been posted but not yet sent to the printer are termed as "Expected."

Our abstracts do not include a keyword section or subject classification numbers. Major organizations that work with these data, like MathReviews, ISI, and Zentralblatt, have to extract them from our document files that they can download anytime free of charge.

When we designed our website, we had mathematicians in mind and not organizations that provide content and search facilities for libraries. It has been my experience that such organizations want publishers to submit to them abstracts and metadata in their own format resembling structured library cards. So far, HJM has ignored such requests. Moreover, according to a poll that I conducted among HJM authors, for work-related search, 55% used MathReviews and Zentralblatt most of the time, 35% Google, and nearly 10% arXiv. There was no interest whatsoever in search facilities provided by Project Euclid, SwetsWise, EJS, or major publishers. Of course, in fields other than mathematics, outcomes of such polls might have been rather different.

Most commercial journals, but also some independent ones, have their document files registered with CrossRef in order to obtain DataObject Identifiers (DOIs). One of the ideas behind the DOI is that it should replace references to URLs of webpages with stable identifiers. The DOI of a document will remain the same, regardless of changing publishers, broken links, torn-down websites, etc. I have discussed the DOI in more detail in [150]. But I can only reiterate that for journals like HJM there are currently no tangible benefits to join CrossRef in order to obtain DOIs. Our URLs are stable because we are not changing them, and it is unlikely that HJM will ever change ownership. Organizations that cover the research literature, such as MathReviews and Zentralblatt, have been using for decades their own internal identification schemes. Moreover, besides payments to the various organizations behind the DOI (e.g., CrossRef), one must not forget the work involved with creating, proofreading, and registering these numbers. While the DOI eventually might prove to be useful and universally recognized, there is also the real possibility that it may turn out to become obsolete because other organizations, for example those creating Search Engines, might have come up with better, more efficient and cheaper solutions.

Laura N. Gasaway [123, p. 5] has outlined a rather grim possibility for referencing material through DOI: "DOI content providers would not only control the indexing, but also access to the indexing and through the index access to the digital object itself."

Some journals have added to their electronic editions certain features that do not make any sense for print, such as internal hyperlinks, or live external links to referenced websites. On the printed page these links show up in shades of gray, and such PDF files cannot be used for journal printing. Thus one needs two types of PDF files, one for screen view and one for print. The Pacific Journal of Mathematics, for example, went this route.

Although hyperlinks are quite easy to implement, after some trial runs and consultations with authors, HJM refrained from adding web specific features to its PDF files. Posting only one kind of PDF file, namely that used for printing, definitely has its virtues. Of course, one should not forget that a computer-savvy person who prefers to read from screen, can always personalize his view by inserting to the PDF file bookmarks, links, or his own comments. This kind of activity is not unlike adding marginal notes.

Nobody can guarantee that electronic editions will be fully accessible to future generations. When we started our website, we had Project Gutenberg in mind, see [205], and adopted its philosophy as well as a mathematics journal can do. Later important decisions, e.g., concerning archiving older volumes, were guided by the same principle. I will come to this issue later.

3 Pricing Electronic Subscriptions

Some independent journals with full electronic editions charge subscribers only for print, and make full electronic editions freely available to the public. Other journals apply a moving wall and provide free access only after a specified number of years. *Open Access* journals want authors to pay for all publication costs and therefore want to abolish the traditional subscription-based business model. Springer created with its more recent Open Choice a hybrid of a subscription-based and open access operation. For a flat fee, which is currently $3,000, an author can have his or her paper posted on Springer Link, with free access to everybody. But the paper will also appear in print.

Finally, *Pay-Per-View* allows libraries and individuals to download individual articles for a flat fee.

Many newspapers and magazines have free electronic editions. Here advertisements must compensate for lost subscriptions. Research journals, with their very limited readership, are by and large not attractive for advertisers.

Some journals that provide free electronic access have included a warning that free access would cease if there were not enough subscribers or sponsors. And indeed, very recently some major journals, like *Geometry and Topology*, have changed their policies of free electronic access and now charge libraries for electronic subscriptions, or make electronic access dependent on a print subscription.

After considering other options, HJM decided on free electronic access, but only to subscribers of the print edition. That is, we currently do not provide the option of an electronic-only subscription.

Actually, finding the right subscription rate for electronic editions can be quite tricky. It has been suggested that the rate for electronic editions should be set not higher than 80% of the print subscription rate. While this sounds quite reasonable, one must not forget that printing costs depend mostly on the number of printed pages, and not so much on the number of copies. Publishers pay, in essence, only for the first 200 copies or so. It does not make much difference in printing costs whether a publisher orders 300 copies or 1,000 copies. But there is a substantial difference between an issue with 250 pages and one with 300 pages. Taking this into account, a somewhat paradoxical situation arises: regardless of whether we are talking about a small independent publisher, or a large commercial organization, any non-trivial discount for electronic subscriptions will cause a financial shortfall.

Here is a quick calculation based on a ten percent discount for the electronic edition. For commercial journals, printing costs are quite low relative to the total subscription revenue. For a typical larger commercial journal that contains about 2,000 pages per year, and has about 800 subscribers, the annual printing costs are about $40,000. Let us further assume that the annual subscription rate is $1,000; that is a very reasonable $.50 per page. Then printing costs for this journal make up only five percent of the total subscription revenue. That is, $760,000 remain for other expenditures and profit.

Now, if this journal offered electronic subscriptions for the reduced rate of $900, then if 400 subscribers took advantage of the lower electronic rate, the subscription revenue would fall from $800,000 to $760,000.[1] Moreover, the costs for providing print copies to subscribers would not change much; they would go down by about $4,000. That is, the journal would experience a shortfall of $36,000; that is with the total costs of printing. The situation would only get worse with any further increase in the number of electronic subscriptions.

For independent journals, printing costs are a big-ticket item. For such journals, about 50% of the subscription income is needed to cover printing costs. But again, a lower demand for print copies will not have much effect on printing expenses. Only when print copies totally disappear, then, unlike their commercial counterparts, independent journals would experience a substantial reduction in their operation costs.

As I see it, as long as there is demand for print copies, offering free online access as part of a print subscription is not only a convenient solution, but it even makes good business sense.[2]

---

[1]We assume here that the total number of subscriptions will remain the same.

[2]The American Mathematical Society, as well as the Society for Industrial and Applied Mathematics, provide free electronic access for subscribers of print editions. However, *Indiana University Mathematics Journal* offers a 20% discount for electronic-only sub-

Besides regular subscriptions, HJM offers the Pay-Per-View option. As expected, see my article [150], it has not created much business interest so far.

Nobody knows whether or when print copies will disappear. There might be an intermediate period during which libraries will keep printed copies for a limited time, but no longer go through the expensive process of archiving.

At any rate, electronic subscriptions have already become increasingly important; currently about 50% of our subscribers have online access. For new subscribers, online access is especially attractive because it also gives them access to all previously published issues, that is, a complete run of the journal. Naturally, most of our new subscribers want to have online access included.

4 Providing Restricted Access

After HJM decided to move up to a "print plus electronic" journal, several technical issues concerning restricted access had to be resolved. I was told by our IT staff that they could accommodate any possibility (e.g., access by IP address, username and password, cookies). It did not take me long to realize that access by IP addresses would provide the most convenient route for a journal like HJM. But what should constitute "access"? Should access be restricted to the issues of a current subscription or should it include previous issues? Should access cease with cancellation or should it remain for all those subscriptions which a library had paid for in previous years? Is a subscribing library entitled to download in a systematic fashion every file it has access to or does the library only act as an agent for its faculty? In other words, does the library buy anything of material value together with access?

In order to tackle in a realistic way this large array of possibilities, one first has to have some idea how restricted access by IP numbers actually works. In principle, this is a straightforward process in which only some details depend on the server and its software. The University of Houston mathematics department is on Linux and uses the Apache.

---

scriptions, claiming that they are following guidelines of the American Library Association. In contrast, *Duke Mathematical Journal* charges $1,520 for print and $1,515 for electronic, that is, print and electronic editions differ by less than one percent. However, for "print plus electronic" Duke charges $1,685, which is a surcharge of about 11% compared to print alone. Commercial publishers have developed all sorts of pricing schemes. Springer has been especially creative, offering "enhanced" versions for print as well as for electronic subscriptions.

As one might expect, all files for which access is restricted are in one directory, called "restricted." This directory can have any number of subdirectories together with its folders. Access is governed by a particular type of file, called `.htaccess`. These files contain registered IP numbers or domains. It became obvious to me that HJM can maintain only one of such access-controlling files. Thus, a request for a file to which access is restricted is honored if it comes from a computer with an IP number that is listed in the `.htaccess` file.

This technical decision has as the immediate consequence that access is granted only to current subscribers. With cancellation, all privileges are lost. I had discussed this delicate issue with various librarians, and the consensus was that our cancellation policy constituted the most reasonable approach: a subscribing library gets print copies for indefinite use while electronic access should be considered a bonus included with a print subscription. As one librarian put it: "With the print copies, the library got what it had paid for, and if it cancels it has nothing to complain about."

It is now commonly understood that access to digital material does not provide ownership of anything, but buying digital material in the form of CDs could be a different matter. However, it looks like libraries are not particularly interested in buying electronic material. And major publishers are not eager to sell.[3] At any rate, the idea of digital libraries has more or less evaporated. Publishers (or in some cases their vendors) provide access to digital material but do not sell content.

If a library wants online access then it has to sign a license. A license is a contract and once signed, both parties must adhere to its terms. Carrie Russel, the Copyright Specialist with The Washington Office of the American Libraries Association (ALA), spelled out the situation quite clearly, see [223]: "Licenses are private contracts between two parties.....Once a license agreement is signed, the agreement takes precedence over any rights libraries or users may enjoy under the federal copyright law." I believe that this applies to current as well as future copyright laws. In other words, an agreement is an agreement.

[3] *Duke Mathematical Journal* has made its first 100 volumes available for sale. A library can either "rent" access for $250 per year or outright purchase the whole collection for $4,000. An additional $1,000 would buy a maintenance contract for 20 years. Thus renting and buying amount to the same sum of $5,000 for the first 20 years; certainly not an incentive to buy.

While the HJM license addresses all the usual points, e.g., authorized users, permitted sites, disclaimer of warranties, of real importance are only two clauses:

(1) The license holder must inform HJM about cancellation in order to have its access numbers removed.

(2) Files cannot be used for Inter Library Loans (ILLs).

(1) assumes that a library operates under some sort of honor code. A violation of (1) may be considered as theft of service.

Allowing ILLs for digital material is a hotly disputed issue. Of course, violations of (2) are nearly impossible to trace but certainly quite common. A remarkably small number of librarians wanted to negotiate our ILL restriction, that is, the copyright clause. A few librarians called me, only to make sure that they understood our copyright restriction. In all such cases, after some discussions, they had no problems with our exclusion of ILLs.

I consider the Pay-Per-View option as a fast and convenient alternative to a slow and potentially illegal ILL process. Our charges are low, $10 per article for individuals, which is close to what "lending" libraries charge for fulfillment of an ILL request.

So far, there has been only one case where a library asked for online access, but an agreement could not be reached. In this case, the library faxed an amendment that, among other provisions, would have allowed it to sell files for ILLs. I showed no sympathy for that, especially because the request came from one of the largest libraries in the country that also has been mentioned as one of the few Digital Mathematics World Libraries. Usually, libraries that argue that they should be entitled to use files for ILLs argue that if they use a printout of a file, then as to that paper copy, existing copyright laws apply, especially when they agree to use for transmission only paper-based fax technology.

As we all understand it, a loan is a transaction that consists of lending and of borrowing, and the borrowed object has to be returned to the lender. Loans of books and paintings work that way. It is also understood that an object on loan is not available by the lender. ILLs deviated over the years from these basic principles, but various Copyright Acts, e.g., USC 108,109, granted libraries exemptions. But as long as copying and faxing is done from hardcopies, the lending party experiences at least some inconveniences, for example, copies have to be found and re-shelved. If ILLs are allowed access to downloaded files, then a click of the mouse is all that is needed. Thus, some libraries have come up with the idea that files should be allowed for ILLs if they are used only indirectly, namely, to provide printouts of requested material.

Then these printouts should be faxed but afterwards destroyed. HJM certainly disagrees with this line of thought. Making the ILL process artificially complicated does not make it legal.

Of course, we honor any reasonable request for exemptions, like a more flexible definition of authorized user and of permitted sites. We also allow IP numbers of proxy servers. But by and large, we have developed very much a "take it or leave it" attitude. We charge all libraries the same rate, and the same concise license should apply to all.

Our license has been scrutinized by dozens of libraries and their legal departments and has undergone quite a bit of polishing over the years. Our license has been largely influenced by the "Terms and Conditions" of the "Online License Agreement Form" used for AMS journals. In contrast, I disliked the "Standard License Agreement" promoted by Yale University Library, see [278], mainly because of its Inter Library Loan clause.

As I have already mentioned, we do not ask subscribers to renew online access on an annual basis. Only cancellations have to be reported. Naturally, there have been violations. If a violation has occurred that warrants termination of access, like illegally maintaining access, then the dean of the library will receive a letter from HJM approved by the UH legal department, stating that online access has been terminated and that a future subscription no longer entitles the library to online access; such access would be subject to further negotiations. If applicable, a copy of the termination letter is also sent to the library's subscription agency. Some violations have been quite blatant, such as cancelling the subscription shortly after online access had been established. Of course, HJM is not the only publisher that has seen dishonest behavior by certain libraries. There is not much a small publisher can do, but I always inform the board of editors about terminations, so the word might spread.

HJM gets nearly all of its subscriptions through agents. Because online access is based on a separate contract between subscriber and publisher, not all agencies are particularly interested in promoting electronic access. Swets has been the exception. But online access is becoming more popular and agencies are increasingly getting involved.

Currently the two largest agencies, Swets and Ebsco, are already content vendors for a number of publishers. However, I prefer a system in which agencies could sell and control access, but publishers would share content only with a third party, preferably a nonprofit organization. For electronic subscriptions, this kind of business model would be the most natural.

5 Adding an Archive

Scanning paper documents to obtain files has opened the door for digitized archiving. While in theory, libraries could digitize their own periodicals, they certainly would face legal challenges if files were made available to other institutions. Only copyright owners are entitled to digitize paper documents and use files without any restrictions. For practical and economical reasons, some larger publishers have commissioned organizations like JSTOR to scan documents and distribute files.

From a business point of view, the opportunity to combine current subscriptions with access to backfiles should give publishers enough reason to do their own archiving. It should be obvious that for a library starting a new subscription, the subscription is more valuable if it includes the "full" journal, and not one containing only current issues. Only publishers who own their archives can offer this new type of complete subscription.

One might argue that publishers cannot be trusted to preserve and maintain the research literature, and that archiving should be left to libraries or to nonprofit organizations like JSTOR. But this approach will, or has already, created monopolies of content holders. Librarians have told me that subscriptions to JSTOR are not exactly cheap, and some recognized world libraries became entangled in lawsuits because of alleged copyright infringements or unauthorized sale of copyrighted material.

While contemplating a few other options, HJM decided that it should pursue its own digitizing project consisting of the first 24 volumes. To obtain bids, I contacted several companies with experience in the commercial, as well as academic, sectors. Generally speaking, digitizing consists of two parts—the physical act of scanning printed pages and the management of the resulting files. Because HJM already had a complete index of all published issues on its website, I expected the digitizer to utilize the data provided by the index for naming and hyperlinking. I decided to choose a company, Princeton Imaging, that had carefully analyzed our website and had offered a bid that included creation of tables of content according to our style, together with hyperlinks to anticipated directories on our web server.

To implement the scanning process, we had to make a few choices. With respect to choosing a file format, PDF was the only serious contender. Incidentally, our digitizer is a firm believer in DjVu, but because we have a relatively small project with no color graphics, Princeton Imaging recommended PDF as the better format. With respect to resolution, we decided on 400 DPI as the best compromise between file size and desired screen/print quality.

Our archived issues are not more than scanned copies of the paper originals; we wanted them to be exact digital replicas of the originals. While our PDF files allow for OCR-based search, nothing has been added. For academic reasons, titles and bibliographical entries of current as well as archived editions are not linked to MathReviews or Zentralblatt.[4]

I do not think that moving walls are a good idea. Especially for the very inexpensive journals like HJM, it makes more sense to include, with a current subscription, access to all backfiles. However, as a service to the mathematical community, all special issues and surveys have been made freely available.

We no longer maintain a large inventory of hard copies. Only a few copies of every issue are kept. For digital holdings, it has become increasingly important to have them stored at different locations. Currently, HJM is mirrored by the mathematics department of the University of Zürich. Besides diverting Internet traffic, this mirror site safeguards our files in case of catastrophic events.

There has been very little interest in purchasing hard copies of our special issues, even of the very important and highly regarded ones. This seems to be a clear indication that if an issue is (freely available) in electronic form, not much demand is left for obtaining hard copies. For periodicals in general, one might conclude that in the future, electronic editions, and not print, will be considered as the permanent and official versions.

6 Final Remarks

I believe that by and large, technical issues related to electronic publishing and archiving have been resolved. Authors of mathematical articles are exclusively using LaTeX for producing their work, and PDF

[4]The board of editors of HJM agreed unanimously that a publisher should not provide links from papers to reviews. Links from reviews to papers are a different matter and are not objectionable.

has become the defacto standard for online publishing and for digital archiving of any type of documents. For the Internet, HTML is the universal language. HTML is quite unstructured and supports visual but not logical design. Certain variants of HTML are better suited for mathematics and stress the logical structure of a page. However, some of these contenders that are meant to replace HTML are still not supported by all browsers, for example, the widely-used IE; and the success of Google is in part due to the fact that Google does not have to be told that a refrigerator is an appliance or that two is a number. CrossRef wants to see URLs replaced by DOIs. While the DOI is widely accepted by commercial publishers, it also seems to be the case that with a few exceptions, commercial publishers are the only ones using it.

The legal issues concerning electronic publishing are much more complicated. Because digitized information is not covered by current copyright laws, for every new subscription, publishers and subscribers must go through the legal process of agreeing on a license. I think that licenses designed by professional societies are best suited to meet the interests of academic libraries and of nonprofit publishers. Our copyright transfer form does not in any way impede the author's right to promote his or her own work, e.g., by posting files on the author's homepage or on arXiv, or by sending files to colleagues. Transfer of copyright to HJM means that the author agrees that only HJM can distribute the author's work actively in a systematic fashion, e.g., through sales to libraries. Our Online Subscription Agreement continues in the same spirit. The license does not impose any restrictions that could cause a conflict for an authorized user who uses HJM files for his or her work, but libraries can download files only on behalf of authorized users and not to serve the interests of clients of other libraries.

HJM had no difficulty with creating its own archive. Thus, we are able to offer electronic subscriptions with full access to all previously published volumes.

There are a number of reasons why I do not support the idea of moving walls. Moving walls create additional work on our side but serve only the interests of nonsubscribing libraries. One also should not forget that freely available material is not entirely protected. Services may pick up free material and bundle it up for sale. I have seen whole journal issues offered by an agency in India. It seems that this is legal because a service has been sold, and not freely available content. I suppose that publishers that provide older issues for free had a service to the public in mind, and not the creation of shady business.

It has been suggested that independent journals should band together in order to avoid being crushed by commercial powerhouses. Unfortunately, all proposals that HJM has received so far did not contain a clearly defined, or attractive, business plan. On the other hand, HJM is very open to the idea of granting to an organization, for a fixed lump sum, the right to sell electronic access to any number of customers it could find. I believe that such an organization should not be allowed to buy content, but only the right to sell and serve an unlimited number of electronic subscriptions. It has been said that whoever holds the content becomes the publisher. Independent publishers must be very careful about joining larger establishments, whether commercial or nonprofit, otherwise they run the risk of being placed under guardianship; and then they might suffer from bad decisions into which they had no input.

I have always been very optimistic about the future of independent journals, especially in this increasingly electronic era. There are many reasons for me to stay optimistic. However, there are also strong negative forces. But let's first set forth some positive facts and trends.

The present state of affairs is promising for independent journals. Like HJM, most academic journals have seen an explosive growth in the number of published articles. In the past, researchers had only limited information about the editors, the scope, and the content of smaller journals. The Internet has changed that. Now everybody operates in the open and independent journals have become more visible.

Independent journals are very inexpensive, and are not under pressure to increase annual subscription rates just because investors are expecting higher profit margins. Independent journals that are published within academic departments increase their rates only to offset inflationary pressures, primarily of nonnegotiable costs like printing and mailing, and not so much for pay increases for their small staffs. Thus, independent journals should experience an ever-increasing price advantage.

Publishing is not high-tech; no publisher has to maintain an R&D department. This was certainly the case in the old days; I am not aware that publishers have been involved in the development of printing technologies. The digital era certainly did not change the situation. In the old days, only commercial publishers could afford expensive and difficult to use typesetting equipment; independent academic journals relied on electric typewriters with exchangeable ballheads, and it showed. Because of LaTeX, now all mathematics journals have the same professional look, and authors have done the major part of the typesetting work in all of them. Moreover, most LaTeX implementations allow for direct compilation into PDF files that are the basis for printing and web posting. Therefore, "Where does all the money go?" This question was raised in 1999 by Michael Barr [24], who contemplated about the high prices of commercial journals. We still do not know where the money is going, but we know for sure that more money is flowing into the pockets of fewer mega-publishers.

High prices of commercial journals are probably the strongest enemy of independent and societal journals. Through highly inflationary behavior, commercial publishers have been trying to claim an ever-increasing share of library budgets. A few years ago, some highly regarded libraries started cancelling a large number of their Elsevier holdings in order to save subscriptions for societal and independent publishers; see the widely circulated article in *Nature* [159]. While not all libraries followed this trend, some certainly did.

In order to distinguish themselves and justify their high prices, commercial journals stress impact factor, membership to CrossRef, and, more recently, usage statistics.

Advertising the impact factor might actually backfire. It is, for most mathematics journals, not impressive anyway and it can fluctuate a lot. Quite recently, various mathematical societies within the EMS have been openly critical about the SCI and the Impact Factor; see the illuminating open letter [213] by the Royal Flemish Academy et al. Incidentally, the impact factor of HJM has seen a substantial increase over the years but it would be unwise to make too much out of this.

"Usage Statistics" have become the latest craze. In a forthcoming article by John Ewing [101], the AMS will express its concerns about the misuse of impact factor and usage statistics. Whenever librarians are contacting me about usage statistics, I tell them that they should ask their mathematics faculty about our journal.

The situation of high journal prices is considered in some circles as so serious that some people want to see the subscription-based business model abandoned and replaced by Open Access. Michael Held [245] and the group behind Open Access Now (see for example [270]) represent opposite sides of this highly controversial movement. I think that Open Access is an extreme solution to a crisis that has been caused by a few commercial publishers. I believe that free market forces will eventually provide a better solution. It looks like some commercial upstarts, like Hindawi and International Press, have already been quite successful.

For all publishers, the illegal copying of electronic material is of greatest concern. This is because we are dealing with a limited market for products in which the price is essentially inversely proportional to the number of buyers. Whether the current situation justifies Patricia Schroeder's rather sarcastic statements about librarians, cf. [171], is a different matter. Of course, Open Access would alleviate the situation, but would also create a very large number of, perhaps, much more serious problems.

Journals have to be produced individually under circumstances in which different issues have very little in common with each other, and where not much in the production process can be automated. Thus, for a publisher, producing more titles does not necessarily lower the production costs for any particular title. Computers and the Internet have made it possible to produce journals of highest technical quality with minimal expenditures of space and personnel. This fundamental fact concerning journal production is probably one of the the main reasons that small academic publishers have become tremendously competitive, a fact that is probably not going to change anytime soon.

Toward a Digital Mathematics Library?

A French Pedestrian Overview

T. Bouche

After an overview of the worldwide mathematical journals ecosystem, we summarize some of the hopes and fears raised by the digital environment. We then review the underlying principles and main features of some of the projects launched by Cellule MathDoc that aim at settling robust foundations and giving wider access to academic mathematical research journals. In turn, they could prefigure some building blocks of the digital mathematics library, which is still to come into being.

1 Introduction

We are living in a period of transition concerning the diffusion of knowledge and the results of research. Printed paper form has been the vector, and to a certain extent the engine, of culture and scholarship during the past 500 years. This paradigm today is largely challenged by new means of production, diffusion, and conservation of the texts. While it is impossible to predict the future of scientific publishing, one must be aware that search for documentation will be made from now on mainly on the Internet, and that access to reference works in a single click is an immense boost for their diffusion across the whole world. Surely, predictions about the death of the Gutenberg galaxy were a little hasty, as there is no better device to disseminate consolidated knowledge today than printed and bound books. Nevertheless, one has to admit that contemporary research is done primarily through electronic means (email, exchange of preprints over the networks, online bibliographic searches, access to the articles themselves through electronic publishing services, etc.). In order for the fundamental heritage that our forefathers left us not to be simply lost for generations to come because of those ongoing changes, it was necessary to undertake conversion from the existing paper literature into a sufficiently rich digital format to allow whatever use might seem appropriate to our descendants. Moreover, it is essential to think globally about this move; that is, let us try to avoid too specific designs in our applications so that they can eventually merge together within the current digital environment and with future developments. We consider that it is our generation's responsibility to initiate this task because it will probably be the last generation that will have been familiar at the same time with the old medium and the new ones.

2 A Glance at the Mathematics Library

> *But the number of the periodical repositories of mathematical literature has become so great, that papers consigned to them, although preserved, as we may hope, for all time, are in imminent danger of passing out of sight within a few years after their first appearance. They are preserved from destruction, but not from oblivion; they share the fate of manuscripts hidden in the archives of some great library from which it is in itself a work of research to disinter them.*
>
> —Henry John Stephen Smith, 1882

2.1 Looking back

Mathematics is special among the sciences in that there is no experiment: the main activities of mathematicians are reading, thinking, discussing, and writing. However, their writings are meant to be a piece in a collective effort to build a logically coherent elaboration from axiomatic foundations to rather sophisticated statements. While any human output is tightly related to previous works in the same field, typically in a dialectic fashion, mathematics might be the only domain where citation is almost never a tool for contradiction. This is why mathematicians usually see mathematical literature as a whole, and would not feel comfortable if some large parts of it got lost, or hidden behind strong barriers. Of course, permanent and immediate full access to all mathematical literature is not required because nice people summarize topics in books, but random access to the actual source is highly appreciated, because it is often illuminating. Reading about the

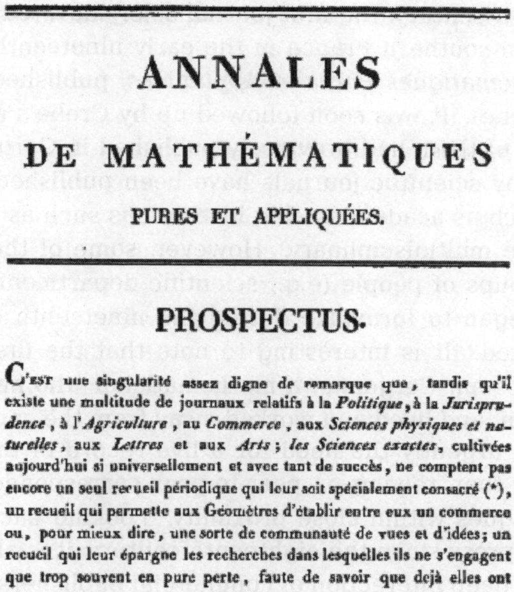

ANNALES

DE MATHÉMATIQUES

PURES ET APPLIQUÉES.

PROSPECTUS.

C'est une singularité assez digne de remarque que, tandis qu'il existe une multitude de journaux relatifs à la *Politique*, à la *Jurisprudence*, à l'*Agriculture*, au *Commerce*, aux *Sciences physiques et naturelles*, aux *Lettres* et aux *Arts*; *les Sciences exactes*, cultivées aujourd'hui si universellement et avec tant de succès, ne comptent pas encore un seul recueil périodique qui leur soit spécialement consacré (*), un recueil qui permette aux Géomètres d'établir entre eux un commerce ou, pour mieux dire, une sorte de communauté de vues et d'idées; un recueil qui leur épargne les recherches dans lesquelles ils ne s'engagent que trop souvent en pure perte, faute de savoir que déjà elles ont

Figure 1. Gergonne's *Annales de mathématiques pures et appliquées* (1810).

invention of a concept in the author's own words yields a much better understanding of the context and intentions behind the discovery. Moreover, as has been already stressed, see e.g., G. Michler [185], it might happen that the proof of an important theorem be so intractable that it never ends up in a textbook, so that resorting to the original article is sometimes mandatory.

Another important aspect of mathematical literature is that mathematics is the common idiom of science, so that virtually any scientist's work relies on some part of the mathematical corpus, but in an asynchronous fashion: while most of current science and technology use well-established mathematics from the twentieth century, specific schools of the past are rediscovered and are an impulse for new trends in some disciplines. It can also be said that many physicists use mathematical theories that have not yet been sorted out by mathematicians themselves.

Later on, we will focus on cumulative and creative mathematical literature where current advances have been appearing for three centuries: scholarly journals. France has a long tradition of producing and publishing mathematics; while the first scholarly journals appeared roughly at the same time in Great Britain and France at the end of the seventeenth century, the first journal solely devoted to mathematics appeared in southern France in the early nineteenth century (*Annales de Mathématiques Pures et Appliquées*, published in Nîmes by Joseph Gergonne). It was soon followed up by Crelle's and Liouville's journals; both of them had previously published in Gergonne's. Since that time, many scientific journals have been published by academic institutions such as academies—big institutions such as Polytechnique in Paris, hence multidisciplinary. However, some of them, published by smaller groups of people (e.g., scientific departments, learned societies that began to formalize during the nineteenth century) were more specialized. It is interesting to note that the first journal was launched by Joseph Gergonne, a mathematician who had failed to be an academician, and lived and worked away from the capital—he probably felt more urgently the need for a live record of the ongoing research, rather than relying on peer-to-peer correspondence or a network of colleagues within close proximity. Looking backwards at the mathematical press since the eighteenth century, it is also interesting to observe the deep interaction of commercial publishers, learned societies, and public institutions in the making of those journals. However, while most editorial boards were acting under the auspices of some trusted moral authority (not always explicitly stated—for instance, the *Bulletin des Sciences Mathématiques et Astronomiques* was launched under the auspices of the École pratique des hautes études, by contract between the ministry and the commercial publisher Gauthier-Villars), mathematicians would not generally supervise themselves the making and marketing of the volumes.

One can wonder about the main benefits these journals have brought to the scholarly community, whether we should list standardization in exposition and style, including notation, organization of the flow of discoveries, or a long-lasting record of the present activity, afterwards resting on library shelves to be easily checked whenever needed. Nevertheless, an important feature of the first journals was to provide bibliographic records of the ongoing publications (books, but translations or abstracts of contemporaneous "competitors" as well). This was then passed along to specific projects like the *Jahrbuch*.

2.2 Current Trends

Today, the general entropy of the system has gone far further, with hundreds of mathematics journals published worldwide, from those published without an economic model by a couple of colleagues, through the many publications of math departments and small societies, up to the big scientific, technical, and medical (STM) publishers with hundreds of titles in their catalogs. Scope, audience, and reputation vary wildly, with no obvious correlation between any of these and price, availability, production quality, etc.

This is a fragile ecosystem that is currently undergoing a deep crisis. The most visible factors are concentration, bundling, and price rises on the one hand, and free access advocacy on the other, which is inspired by the open source movement in the software industry, which is at first glance a surprising paradigm for designing new ways to produce and disseminate intellectual achievements. Over the past decades, it is amazing to observe how often academic publishing has been outsourced when it was once done in-house, and otherwise internalized. The reasons for outsourcing typically are that the scholars feel, at some point in time, that commercial vendors can better accomplish the technical tasks such as proofreading and printing, but also of managing subscriptions, or branding and selling the journals—they are more reliable and recover their higher costs and profits by better marketing. The reasons for internalization are symmetrically based on disappointments with regard to the quality-price ratio of services by external providers compared to what scholars think they can easily do themselves. This is a circular movement that will not soon reach a state of equilibrium. As the then Elsevier Science chairman (now Springer Science+Business Media CEO) Derk Haank [131] puts it: "It is ironic that the whole world is talking about out-sourcing and the academic community would in-source a tedious job like publishing."

One possible answer to this remark is that librarians are not satisfied with the many directions that some publishers opted for recently, thus leading them to undertake themselves some of the tasks traditionally performed by publishers. For instance, publishers are more concerned with their immediate future—the rate of profit they will present to their shareholders at the end of the fiscal year— than with taking care of their unsold items, which are usually sold at paper's value after a short period of time. It is generally considered that it is a library's job to carefully store the publisher's products and index them so that researchers can access them when required. The early years of electronic publishing have placed a certain amount of burden on librarians because the typical offer was a site license that allowed a limited-time access to electronic resources hosted on servers operated by the publishers. This placed the library in a situation of full dependence upon the publisher, as the first mission of libraries—storing intellectual output in order to provide local access to it—was now resized to a minimal one: managing subscriptions and controlling access to third-party services.

While these issues do not look very specific to mathematics, they have a special meaning to the mathematical community because we are probably the scientific community that relies most on its past publications, and we might be the one in the STM area that has kept the largest diversity, where the mainstream publishers do not yet entirely shape the landscape (see some convincing discussion by Pierre Bérard [30]).

It is natural that new ways of producing and disseminating mathematical research should be inquired into as new tools and methods appear, but we must beware that what drives the present choices is mostly vanity and greed. For instance, when the paper industry found cheaper ways of producing something that looked like premium white paper made out of tree trunks rather than softer fibers, publishers adopted quickly what was soon to be known as acid paper, which would start destroying itself later. The use of the Linotype, and then electronic typewriters, in the production of complex mathematical articles was never for the sake of mathematical expressions' clarity or better typography, but of cutting costs (for a nice set of samples, see Donald E. Knuth [160]).

2.3 Looking Forward

Plenty of literature has already been devoted to the future of our academic publication system: we are not going to add yet another prediction to the list. While scholars were arguing, the industry was moving fast. The near future is already largely dictated by the big players, and very large technological platforms have been set up (e.g., Extenza, Highwire press, Ingenta, Metapress, Scitation, CrossRef) and market leaders have assembled gigantic bundles (e.g., ScienceDirect, SpringerLink, JSTOR, AIP) where no access is given for free. Ironically, it is often not even easy to know what content is actually available to subscribers; sometimes even browsing the collections is not allowed. In some instances, messages like "references secured to subscribers" are displayed for papers having no references at all, or for papers that are not yet accessible at the website. But maybe the worst practice (which is becoming routine) is to secure access to texts that are in the public domain.

The only device for interoperability among STM publishers, for the sake of linking, is CrossRef, which provides a general infrastructure based on a central unique registry allowing one to link to some other publisher's website at article level without having to dig into its catalog. In fact, the full catalog is exported and exploited by commercial third parties like generic Internet search engines, those dedicated to the academic literature, or highly integrated tools designed for the rich scholar like Scopus.

Once again, the idiosyncrasies of mathematical literature fall by the wayside of these important maneuvers. Probably the most influential proposal for a more responsible shaping of the digital near future of the mathematics library has been the recommendations by the CEIC of the IMU [199], and this is one of the bases on which we settled when designing the French projects to be discussed below—namely: build on structured open formats, keep all navigational items free, and provide eventual free access to the articles themselves. For a publisher, obeying these recommendations is kind of a bet—offering much to anyone while charging only for a small subset of the features whose development you invested in might raise sympathy, but will it drive revenue?

We are getting to a world where we have a standard implicit commercial practice spread across big publishers, which is to secure everything to subscribers, but to set the subscription fees so that, in fact, anyone *can* be a subscriber, and to acquire such large journal portfolios that research centers cannot bypass them, setting up a situation in which everyone *must* be a subscriber. On the other hand, if the standard practice of less aggressive publishers—maybe those that have a stronger connection to the scholars themselves, that publish few journals, in a restricted disciplinary area—is that recommended by the CEIC, one might wonder how long they will get enough income to survive. Having the researchers archive their preprints themselves is an honest alternative to obtaining access from greedy publishers. But if a publisher takes care of providing a high quality interface, which is not a small investment, including digital backfiles that are freely accessible after some reasonable moving wall, then the only electronic pieces of information that remain to be sold (or exchanged) are the full text of recent articles, which is something one can often do without. This is especially so when the preprint is widely available, or when you can get an "electronic reprint" from the author himself. The main value of the refereed journals system as it has been in existence for a few centuries is the complex process of validation of the sheets sent to the press and the duration of their content in well-curated libraries, so that the texts are safe and well ordered. The traditional paper publishing paradigm assumed that publishers carefully produce the volumes that are bought by libraries in order to be stored. With the vanishing of paper editions, we cannot ask the publishers to take on the tasks traditionally carried out by the librarians, without charging for these new services. We may hope, however, that some digital libraries will devise protocols to transfer newly generated content to their virtual shelves, letting the publishers concentrate on what they do best—choosing content and presenting it in a desirable form.

3 The Digital Mathematics Library

> *With all the money that has been spent on prototypes that aren't working anymore we could have built a simpler but more sustainable digital environment.*
>
> —Martin Grötschel [129]

At the turn of the century, there was a lot of activity that we can now look back on as a "wish for a DML," where DML means digital mathematics library, which in turn had as many meanings as interested parties.

3.1 Beginnings

While some of the protagonists focused primarily on retrodigitization, others felt from the start that what was needed was a comprehensive (thus heterogeneous, searchable, interrelated, and distributed) network of all the existing mathematical literature [79, 184]. Since its inception, however, the DML concept has provoked many discussions, but no significant outcome. To some degree, the initial direction was set by John Ewing's "white paper" [102] as an endeavor to digitize all the past mathematical literature, which required some sort of worldwide cooperation, competition, and organization. So far, some competition, and some cooperation between "competitors," have occurred, but the lack of a real organization, which might be related to the lack of large-scale funding, has stalled the project.

Notice that the amount of valuable mathematics already available digitally is considerable, but all those individual bits do not glue into anything resembling a library. Moreover, as so much of the new additions are driven by commercial appetite, many basic principles are left out. For instance, the Springer "Online Journals Archive" bundle has been assembled according to principles that make sense commercially, but not scientifically, at least for mathematics: the included journals

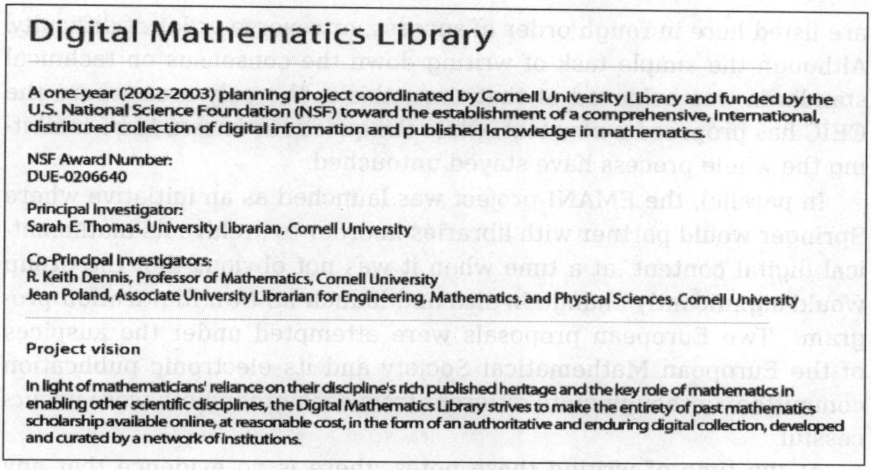
Digital Mathematics Library

A one-year (2002-2003) planning project coordinated by Cornell University Library and funded by the U.S. National Science Foundation (NSF) toward the establishment of a comprehensive, international, distributed collection of digital information and published knowledge in mathematics.

NSF Award Number:
DUE-0206640

Principal Investigator:
Sarah E. Thomas, University Librarian, Cornell University

Co-Principal Investigators:
R. Keith Dennis, Professor of Mathematics, Cornell University
Jean Poland, Associate University Librarian for Engineering, Mathematics, and Physical Sciences, Cornell University

Project vision

In light of mathematicians' reliance on their discipline's rich published heritage and the key role of mathematics in enabling other scientific disciplines, the Digital Mathematics Library strives to make the entirety of past mathematics scholarship available online, at reasonable cost, in the form of an authoritative and enduring digital collection, developed and curated by a network of institutions.

Figure 2. Home page of the DML project at Cornell.

are those in the Springer group that are written today, mostly in English. They are scanned at 300 dpi and OCRed using an English-only OCR (which means, for instance, that the French letter *é* can be recognized as a 6, yielding such poetic statements as "D'aprbs la d6finition de la d6riv6e g6om4trique"). We might be concerned by the fact that only the latest series of the Italian journal *Annali di Matematica Pura ed Applicata* was recognized to meet those selection criteria, thus we will have to look elsewhere for the seminal works of Beltrami, Betti, Cremona, etc. The actual state of disorganization is fairly well depicted by Eugénio M. Rocha and José F. Rodrigues in [217]; see their chapter in this volume for an update.

To date, the DML has been

- an informal club of enthusiasts that met at San Diego, California, in January 2002;
- an NSF planning-grant hosted at the Cornell University library (2002–2004);
- an IMU project coordinated by the CEIC through its WDML subcommittee that was formed by IMU's president during the summer of 2003, but never started working;
- a wdml.org website with some sketchy information provided by the CEIC since 2003.

The second event was the most productive, see [85]. Six working groups were formed, addressing the following issues: economic model, archiving, metadata, content, rights, and technical standards. They are listed here in rough order of success, or reverse order of difficulty. Although the simple task of writing down the consensus on technical standards was achieved at that time, this is the only area where the CEIC has proposed a new text, while the questions known to be inhibiting the whole process have stayed untouched.

In parallel, the EMANI project was launched as an initiative where Springer would partner with libraries in order to archive its mathematical digital content, at a time when it was not obvious that the group would significantly change in size and launch its own digitization program. Two European proposals were attempted under the auspices of the European Mathematical Society and its electronic publication committee—although quite different from each other, both were unsuccessful.

At the time of writing these notes, there is no evidence that any DML-related international effort will ever succeed. The recent history shows that the stakeholders are eager to share their expertise and discuss the issues, but that no trusted entity has yet emerged in favor of which they would agree to lose some control over their collections.

3.2 Retrodigitization and Archiving

It is a strange fact of contemporary life that, while everyone faces problems reading electronic documents a couple of years old (or even just finding one that has been stored on some local drive), but enjoys reading books that are decades old, we are convinced that paper will soon vanish and will have to be replaced by some electronic device. In fact, we even believe that we are going to learn from our bad experience so that we will come up with very good tools for long-term preservation of electronic files. There is no evidence so far that supports this optimism, but we will have to wait some time to know!

However, it seems obvious to anyone that mathematical literature will be more useful once it has been converted into a suitable digital format, especially if we succeed in converting the legacy article stock into an active web of references. The first action should thus be to digitize the existing literature for which no satisfactory digital surrogate is already available. The next obvious actions are to store the newly created files in some virtual library system, to build multiple preservation and access services on top of that, and to set up a mechanism so that the ongoing production ends up in the same system.

As for long-term preservation, one can share the skepticism of Christian Rossi [221], namely that the more meaning or structure you store in a file, the more likely it is that you lose the environment or documentation that allows you to exploit its content (note, e.g., that the TEXBook is a TEX program, that the PDF format is described in a PDF file, that a very well-structured XML file does not bear by itself any information about its own structure, its meaning, and its expected usage). So all we can hope for is that very clever people will devise high-end systems that will enable us to store safely all relevant data and metadata together, in some format so simple that future generations will not have difficulties reverse-engineering it, and on some medium that will not be already unreadable when their curators will learn that it has become obsolete.

In fact, one lesson of successful digitization projects might be that it is possible to build relatively easily, and at a relatively low cost, a full-featured electronic edition of a journal, starting from paper volumes. Due to this observation, together with the difficulty of reuse of sophisticated digital formats out of their context, we are inclined to think that the safest digital archive of a paper-based publication (but maybe of a digitally created one as well) is a flat directory of bitmapped pages in some straightforward uncompressed format.

3.3 Born Digital Content and Integration

The questions regarding long-term archiving of electronic journals, secured away from publishers' fate, have yet to be answered: some concurrent proposals are on their way, many of them being held by organizations that are not necessarily economically secure in the long term.

We are also concerned by advances in digital rights management (DRM) technologies together with laws enforcing full control to the rights' owner over the usage restrictions at the time of publishing. This might lead to the massive dissemination of unarchivable data. For instance, an existing DRM technology (protected PDF, by Vitrium Systems) claims that it enables a publisher to produce apparently benign PDF files, which you can print a given number of times, whose use is monitored by the publisher, whose content can be modified at any time—updated or merely removed. All these features, thus the very possibility to access the published text, rely on a quite volatile state of the art—current software and networking technology—and quite a few commercial partners, some of which are start-ups!

We could consider that the paper version is a better authority for long-term archiving, but we should realize that current paper printouts are paper surrogates of electronic systems, no longer the other way round. For instance, given the low diffusion of specialized mathematical journals, many publishers will provide a paper edition that is made using print-on-demand systems. But, while liquid ink chemically alters the paper that absorbs it and insures maximum longevity for the text (in some cases, the text has been observed to be the only lasting piece of some books printed on acid paper!), laser "printing" consists in cooking some rigid polymer over the surface of the paper which will eventually reject it.

Digitization might be the only way to secure an archive independent from the possibly bad choices of today, still for quite some time ahead ...

3.4 Concluding Remarks

Currently, the DML is an idea with great potential that drives some activity, but most actions are done in competition and independently, with very short-term outcomes in mind. The best that we can hope for is that we will still have somewhere, in some usable format, the mathematically valid content that is output nowadays, and that we will not lose the content that is already at rest on our library shelves.

An effective digital mathematics library should address the fundamental functions of a library: archiving (for access and for preservation) and acquisition. The first issue, electronic access, has been rather well understood and is already implemented in a quite satisfactory manner in existing retrodigitization programs—however, the long-term sustainability of these programs is unknown. The second one, digital preservation, is, and will remain, a vastly open question. Indeed, it is only in case of failure, in the future, of the current preservation plans that we shall know for sure whether they were robust enough to succeed. The last one, acquisition of new electronic content and its transfer to virtual shelves for access and preservation, is in a very problematic state. Some promising projects such as Cornell University library's Euclid do provide the feature, but they typically go as far as hosting the online edition of the journals in order to be able to get enough data to archive them; this gives no hint on how we could convince commercial publishers to feed a stable and ever-growing DML.

Nevertheless, we hope it will be useful to draw a sketch of the picture we envision, and to detail the implementation of a prefiguration of some of its components in the very specific French context.

4 The French Actions

What is important:

- *Make things easily usable, even better, make them simple!*
- *Long-term funding and not short-term projects.*
- *Sustainable activities are more important than hyped killer apps.*
- *International and interdisciplinary cooperation.*

—Martin Grötschel [129]

4.1 The French Ecosystem

We have in France a long tradition of mathematical publishing, which yields the current situation in which about 20 journals are currently published. We shall distinguish the categories of these journals:

1. independent journals (usually published by a university department): *Annales de l'Institut Fourier, Annales Mathématiques Blaise-Pascal, Annales de la Faculté des Sciences de Toulouse, Journal de Théorie des Nombres de Bordeaux, Cahiers de Topologie et Géométrie Différentielle Catégoriques;*
2. learned societies publications: *Bulletin et Mémoires de la SMF, Astérisque, Revue d'Histoire des Mathématiques* (SMF), *Revue de Statistiques Appliquées, Journal de la SFdS* (SFdS);
3. journals published by a commercial publisher under the auspices of an academic institution or a learned society that owns the title: *Annales de l'Institut Henri Poincaré* (IHP/Elsevier), *Annales Scientifiques de l'École Normale Supérieure* (ENS/Elsevier), *ESAIM–Control, Optimization and Calculus of variations, ESAIM–Mathematical modelling and Numerical analysis, ESAIM–Probability and Statistics, RAIRO: RO* (SMAI/EDPS), *Journal of the Institute of Mathematics of Jussieu* (IMJ/CUP), *Publications Mathématiques de l'IHÉS* (IHÉS/Springer);
4. the unique case of a public service operated by contract by a commercial publisher: *Comptes Rendus de l'Académie des Sciences, Série A Mathématiques* (Académie/Elsevier);
5. journal titles currently owned by a commercial publisher: *Bulletin des Sciences Mathématiques, Journal de Mathématiques Pures et Appliquées* (Elsevier).

To my knowledge, most journals in the first three categories accept exchanges. It might be useful to note that one finds the most scientifically respected journals under category 3, which is also the category with the most journals, while each category holds one or more internationally respected journal; scientific reputation is rather independent of the organizational structure. Most of these journals are supported by the CNRS, which is a public institution of the French state (the support might be to provide grants as well as permanent technical staff, employed by the CNRS, to the journal).

4.2 Digitization

In France, in our context, digitization is spelled NUMDAM (aka *numérisation de documents anciens mathématiques:* old mathematical documents digitization). This might change, as some publishers and librarians could decide to digitize their collections themselves, but this is not expected soon. We could also mention projects that were not designed specifically for mathematical journals but deal with some of them, like Gallica, or a project that is similar to (in fact partly inspired by) NUMDAM in the fields of social sciences and humanities: PERSÉE. What all these programs have in common is that they are managed by public institutions of the French republic, and meant as a public service, hence most of the collections are freely accessible.

Outline. NUMDAM was initiated just before the year 2000 by MathDoc's directors of the time, Pierre Bérard and Laurent Guillopé, thanks to generous support from the ministry of research. The *Cellule de Coordination Documentaire Nationale Pour les Mathématiques* (aka MathDoc) is a small service unit belonging to both the CNRS and the University Joseph-Fourier, located in Grenoble. It is dedicated to providing bibliographical services at a national level to the mathematical community, which is why it is headed by mathematicians, and employs a balanced staff of librarians and computer scientists.

The actual industrial production is outsourced: MathDoc concentrates on inspecting and preparing the collections, controlling the output of the production, archiving, and maintaining the online access at NUMDAM website.

We have only consumed public funds so far, adding to our initial funding specific funds from various sources. Although we consider ourselves a part of the DML, we would like to emphasize that MathDoc is not a library, and NUMDAM is not associated with any library in particular; MathDoc owns no documents itself. NUMDAM is meant as a service to other parties, under the auspices of the French mathematical community. Similarly, NUMDAM posting is not considered a new edition—just the exact copy of the paper edition, in another medium. MathDoc acts as an agent for existing publishers; it is not a new one.

NUMDAM's design was a long task, and we must acknowledge the support that we received from our funding partners, who did not pressure us to produce something quickly, but left us to think a while on how to do it best. The first online posting did not occur until December 2002, when two journals appeared. At the time of writing, five years later, we offer 20 serials, 1 series of proceedings and 27 important seminars. We first started with the obvious, highly regarded, and currently publishing, pure mathematical journals, but we now provide a much more comprehensive collection, with journals in applied mathematics, mathematical physics, and statistics, consisting of more than 26,000 articles spanning over 560,000 pages. These include the first ever mathematics journal (*Annales de Gergonne*, 1810–1831) and our first non-French participating journal (*Compositio Mathematica*), thanks to an agreement with the Foundation Compositio and the London Mathematical Society.

The recent additions concern one French journal at the boundary of mathematics and social sciences (*Mathématiques et Sciences Humaines*), two Italian journals (*Annali Della Classe di Scienze, Scuola Normale Superiore di Pisa*; *Rendiconti del Seminario Matematico della Università di Padova*), seminar proceedings from the École polytechnique (around Laurent Schwartz) and Collège de France (around Jean Leray), and the Bourbaki seminar. As the collections keep growing steadily, we should reach the mark of one million digitised pages at the end of the current phase of the program.

Goals. The main concepts that led to the decision of raising funds for the digitization of mathematical literature published in France were: long-term archiving, visibility, and availability.

Nevertheless, the inception of NUMDAM had its roots in ongoing discussions at French and European levels with inspiration from JSTOR, CEIC, and DIEPER, while Gallica was already viewed as an old project.

As with so many projects in this field, the main motivation was the commercial policy of big publishers: rising prices, bundling, and a licensing model for electronic access that did not make clear what would be available to the average researcher if the publisher was ever to change policy, or to get out of business. So the motivations were very much bound to dated interrogations (see Section 2.2) about big publishers—especially those who are in such a dominant position in their domain that they can rule the market.

One of our goals was thus to build a safe archive of the *French* mathematical cultural heritage, that would be maintained by some public institution, and would not depend on the internal archiving policy of the current publishers, or even on their future destinies. As a side note, we observe that, when some of these publishers decided to digitize their own collections, they had to query libraries because they had no stock and, moreover, much of the collections came from the acquisition of other publishing houses that they had dismantled right away.

The other, companion, goal was to enable independent and society journals to have the same Internet presence, so that they could compete in functionality and expose their rich history, and would not need to compete on scientific excellence alone. This logic has naturally been followed up in the CEDRAM project, which proposes to the same stakeholders to rethink and upgrade their born digital production to current best practices; see Section 4.3.

There is an obvious competitive advantage for visibility in being an early mover. Being more visible surely enhances usage, hence impact. As the medium, and much of the paradigm, move, one can bet on a redistribution of cards. It is thus natural that national funds be spent in digitization as a strategy to get more return on previous investment in this new era.

Finally, we would like to note briefly one issue in our brave new world of scientific publishing: mathematical literature is highly multilingual in essence, and current mathematics is the result of a worldwide effort of thousands of human beings across centuries and boundaries. This is a barrier for the modern mathematician to accessing primary writings in Arabic, Chinese, Greek, Japanese, Russian, or most European languages; however, it is also an opportunity to read some great mathematicians' thoughts in their own words. As of today, the French idiom is the last one, besides English, used for publishing major works in international journals. Each other country with a strong tradition gave up during the last century. As the DML gets populated, it gets more multilingual, and it could reinvigorate linguistic pride, and thus raise higher the quality of the language used in communicating mathematics. We hope that NUMDAM website's bilingualism will sustain the continuing effort of the French community in writing important articles in French.

Copyright model. Our copyright model can be summarized by the following items. The underlying idea is that our work should modify the previous state of affairs as little as possible:

- The electronic version is under the journal title's owner's control (usually an academic institution or a society), as was true for the paper version.
- Authors are asked to transfer their (exclusive) electronic copyright to the journals if they are still alive.
- We contract with the copyright owner to allow online access on the NUMDAM website.
- We define a moving wall on a journal-wide basis (the most frequent term is five years, and there are few free-access journals). We view free access after the moving wall period as a fair counterpart to public funding.

Standards. Because we are not trying to create a new edition, we will not make new editorial choices, thus we respect any aspect of the paper publication: we deal with full journals backruns, and we scan every page at high resolution (600 dpi black and white for text pages, gray or color when appropriate). We retain the original page format in the delivered files, which thus provide a faithful image of the printed material. In addition, we consider that we are responsible for the metadata generated within the project, links, and other edits, hence we keep a clean separation between articles as images and added metadata.

We capture detailed structured metadata allowing one to search basic bibliographical data plus plain-text and cited references. This means that we have a metadata model that allows us to describe journal runs, made of physical (bound) volumes, which are essentially composed of articles. Among the metadata describing an article are its typical bibliographic reference (which is fielded), its abstract, its OCRed full text, and its bibliography, as it is printed, with important bibliographical elements tagged.

We provide a clear identification of the originating journal on all supports through the visible mention of the origin and links to the website of the publisher on the individual article's covers.

Our user interface to the archive is dual—it is made of both the article metadata exposed on our website, and the user version of the scanned article itself. We consider that it is good practice to associate to any article a stable or persistent URL, which should not point to a graphical file, but to the nevralgic center where all the relevant metadata is collected and exposed using a plain HTML format. These are the vertices of the DML graph.

One important reason for this dual design is that metadata features are subject to change, while the article's text is not. Two clear exam-

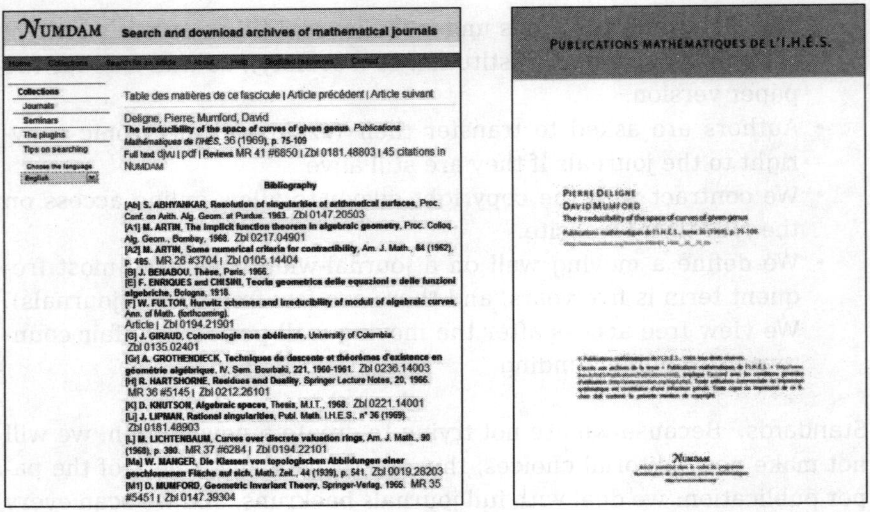

Figure 3. An article from *Publications mathématiques de l'IHÉS* at NUMDAM.

ples are the activation of new citation links as the DML grows, and the correction of errors in metadata.

We have therefore added to each full-text file a first page that is similar to a reprint cover. This provides all the necessary information for users to clarify the origin of the file, the allowed usage, and its permanent URL, which is an actual link from the article file, where its full record is available, with all navigation features up to date.

Links. A hyperlink network places the article in proper context; we add as many meaningful links as possible to each article. For instance, some items in the article's bibliographical reference are clickable, e.g., clicking on "author" gives the list of the author's papers (using a local authority list of authors), clicking on the issue number leads to the table of contents of that issue. There is, of course, a link to the full text (in both PDF and DjVu formats) if freely accessible, or to the article's location at its publisher's website, if applicable. If the article has an erratum that we are aware of, then we provide a link. If the article has been reviewed in one of the reviewing databases that we know of (*Jahrbuch über die Fortschritte der Mathematik* (JFM), *Zentralblatt MATH* (ZM), *Mathematical Reviews* (MR)) and we can match the review there, then we add links as well. On some occasions, we have more special links: one old journal had the special feature that some of its longer articles were spread over various volumes, printed in arbitrary page ranges with no consideration for logic, so that a sentence could be running at the end of a page, and you had to wait for the next volume to know how it would finish, and this resulted in many chunks of the same article in our database—we added links to the list of all the parts from each of them. In the case of the probability seminar of Strasbourg, we had an already-running database of third-party written abstracts of all the talks from the initial period—we linked these abstracts to the texts, and added links from the articles similar to those to other reviewing databases.

But most of our links concern our bibliographical references cited from individual articles; those quotations have been matched to the extent possible to JFM, ZM, and MR reviews. Moreover, when a cited article happens to be part of NUMDAM, we provide the direct link. We started to add similar outer links as a byproduct of the mini-DML that will be discussed below. Without taking into account their respective coverage periods, we have matched 75% of our references to a ZM review, 68% to MR, 4% to JFM, 9% to NUMDAM itself, and 12% to other digitisation centres thanks to the mini-DML facility. This means that we provide around 21% of cited references with a direct link to the cited article's full-text, and still more through the reviewing databases.

Collections. We summarize in Tables 1 and 2 the serials that have been handled by NUMDAM.

Future plans. At the time of writing, we can consider that we have completed our initial task: digitizing important French mathematical journals and making their mathematical content widely available. We have even gone somewhat beyond this goal, as we have been dealing with older material, seminars, and European journals. Our main failure is the fact that the titles currently owned by Elsevier Science, as a result of its acquisition of Gauthier-Villars, have not been digitally archived at all, except for the early years at Gallica, with lower standards than ours. Some other important resources have yet to be digitized because their overcautious owners have not yet set a clear policy regarding archiving or access.

We shall concentrate many of our next efforts on the improvement of the website's features (MathML versions for the OCR and metadata, better full-text search engine—including math expressions, and ranking hits according to relevance), and better integration with other projects. An important feature of the archive is that it should stay alive,

Title	Period	Owner	Volumes	Pages	Articles
Ann. Fac. Sci. Toulouse	1887-2002	University	207	36 270	1 041
Ann. Gergonne†	1810-1831		22	8 700	1 083
Ann. inst. Fourier	1949-2000	Assoc. A.I.F.	156	51 054	1 811
Ann. I.H.P.†	1930-1964	I.H.P.	71	5 860	147
Ann. I.H.P. sér. A, B, C	1964-2000	I.H.P.*	119	55 000	2 400
Ann. math. Blaise-Pascal	1994-2002	Labo/UBP	19	2 554	165
Ann. Sci. École norm. sup.	1864-2000	É.N.S.*	295	68 898	1 867
Ann. univ. Grenoble†	1945-1948	UJF	3	1 006	47
Ann. Scuola Norm. Sup. Pisa	1871-2001	SNS	253	46 062	1 726
Bull. Soc. math. France	1872-2000	S.M.F.	167	45 774	2 608
Mém. Soc. math. France	1964-2000	S.M.F.	134	18 118	396
Compositio Math.	1935-1996	Found. Comp.	275	39 500	1 934
J. théor. nombres Bordeaux	1989-2003	IMB	33	7 700	431
Journées É.D.P.	1974-2003	C.N.R.S.	31	5 976	514
Math. et Sciences humaines	1962-2000	EHESS	148	12 600	763
Publ. math. I.H.É.S.	1959-2000	I.H.É.S.*	92	17 424	344
Rend. Sem. Mat. Padova	1930-2000	U. Pad.	108	33 300	1 980
Rev. Stat. appl.	1953-2004	SFdS	207	22 150	1 676
Sém. Bourbaki	1948-2000	Assoc. N. B.*	45	18 200	920
Sém. Leray†	1961-1977	Coll. de F.	38	3 200	129
Sém. Paris†	1953-1985	Secr. math.	146	37 700	2 647
Sém. Polytechnique	1969-1996	X	37	12 800	910
Sém. Proba Strasbourg	1967-2002	IRMA*	37	17 352	1 254
* Contract with a commercial publisher. † Dead serial.				567 200	26 800

Table 1. Serials currently online.

not only by active maintenance of the whole system, which is something MathDoc is committed to, but also by growing with the addition of the new articles, and this is an area where there is still a lot to be done.

Our policy is to store metadata of post-digitisation articles as soon as they are published, and to post the full texts when they are freed from the moving wall. We implemented this policy with three journals handled by Elsevier (including the *Annales* of ENS and IHP) in 2007,

Title	Period	Owner	Volumes	Pages	Articles
Ann. Fac. sci. Univ. Clermont	1962-1993	UBP	37	5 050	328
Cah. topol. géom. différ. catég.	1966-2006	A. E.	184	16 300	691
Nouvelles annales	1842-1927		94	50 700	8 648
Bull. Darboux	1870-1884		27	9 400	609
SMAI journals (prior to ESAIM series), 1914-1945 French thesis, ...					

Table 2. Forthcoming serials.

and four by CEDRAM. We are willing to update the NUMDAM database with the metadata of newly published articles on a regular basis from all suppliers, adding links from NUMDAM to the actual article's location at its publisher's site, so that the archive also drives audiences to the journals we support. For this to occur, we still need support from more publishers.

MathDoc has had some collaborations, in parallel to NUMDAM:

- The Bourbaki archive project has digitized all the manuscripts, drafts, letters, and internal publications from the early days of the association. We have built a database on top of this, and provide web access for researchers and the curious.
- We are using our scanned articles in an SMF edition of the collected works of Laurent Schwartz, which will be a partnership between SMF, Polytechnique, and MathDoc.

4.3 Redesign of Journal Production

There are many ways to measure whether a journal is successful or not. Some questions to ask are: How many paid subscriptions are there? How many exchanges occur? How much profit is generated? How much subsidy is consumed? How many paper submissions are there? How many papers are published? Among those, how many are good mathematical papers? Would *you* be proud or ashamed to publish there?

If you are an institution providing funding to various journals, you might wonder whether it is your name or logo that gives value to the cover, or the other way round. For instance, small universities tend to support a laboratory's journal and happily accept exchanges from all over the world in the hope that they will increase their visibility. This is also to some degree the rationale that is behind the official support for Open Access by funding agencies worldwide (e.g., Budapest open

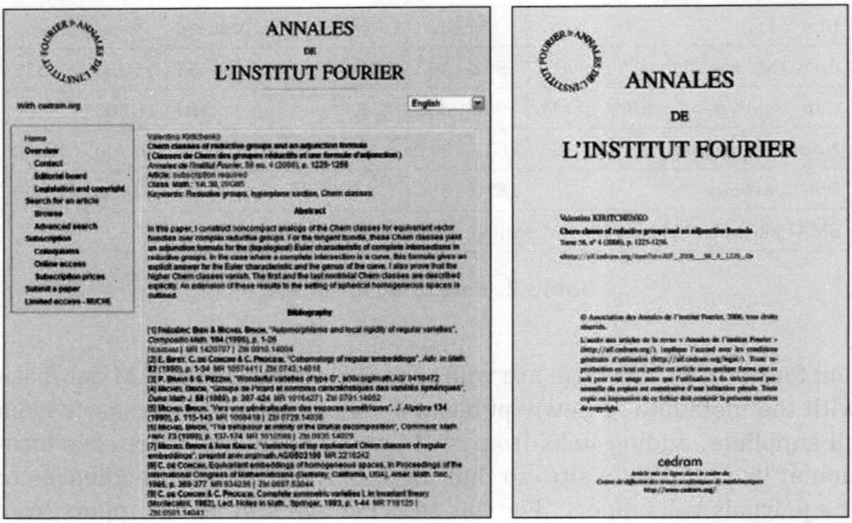

Figure 4. An article from *Annales de l'Institut Fourier* at CEDRAM.

access initiative, Berlin declaration). If you finance research, you prefer it to be visible, used, and referred to, so you prefer the derived publications to be universally accessible.

During the first years of this century, the mathematics division of the CNRS conducted a survey on the mathematics journals it supported, and the results were puzzling with respect to the cost efficiency, regardless of the measure for efficiency that you choose from the above list. We were then in the strange situation where the NUMDAM website was acclaimed for its quality, usefulness, and visibility, where every individual journal's fame and legacy papers added weight to the whole lot; while they were divided on the current front, relying on much amateur or unpaid voluntary work, thus being successful in very different areas. As this is essentially a management issue, this is not related to the quality of published articles. It was then discussed inside the community whether some synergies could be brought out, and this ended up in a seemingly simple motto: "Let's make a French math journals portal!" It was also decided that MathDoc would undertake the effort, Yves Laurent being its director.

This led to the CEDRAM project (an acronym for Centre de Diffusion de Revues Académiques de Mathématiques, meaning something like "Center for the distribution of academic mathematical journals"), which is viewed as a combined effort by the participating journals and

MathDoc, with advising partners, including the French mathematical societies SMF and SMAI, making up a steering committee whose president is Claude Sabbah [41].

Currently, CEDRAM is mostly a set of internal tools meant for the production of mathematical journals with full-featured paper as well as electronic editions, but it is also a platform for hosting their electronic edition. The visible part of the project is at the CEDRAM website, which is the awaited portal: it provides links to all French mathematical journals supported by the CNRS, and a combined search engine covering all the articles of the currently hosted journals.

The main design decision for this platform was that it should provide NUMDAM features at a minimum, plus access control to full texts, and a wide versatility so as to preserve a distinct look and feel for each journal.

Each journal remains completely independent, in terms of scientific content as well as financial balance. The synergies are thus currently limited to sharing a common underlying platform, and being simultaneously present on the portal. More options like bundled subscriptions will probably be introduced in the course of the project.

Journals. As a matter of fact, the founding journal partners of CEDRAM all belonged to the first type of journal listed in Section 4.1. One can even remark that all of them were published by mathematics laboratories from non-Parisian universities. The project was not designed exclusively for them, but it seems that they are those for which there is no doubt that the benefits exceed the burden of conforming to some production standards, and of changing some production steps. This has mildly changed as we help the SMAI (French society of applied mathematics) launch a new journal in 2008: *MathematicS in Action.*

Four journals are online since year 2006, opening the service:

- *Annales de la Faculté des Sciences de Toulouse, Mathématiques*;
- *Annales de l'Institut Fourier* (Grenoble);
- *Annales Mathématiques Blaise Pascal* (Clermont-Ferrand);
- *Journal de Théorie des Nombres de Bordeaux.*

In 2007–2008, updates concern the addition of three series of seminar proceedings, and the aforementioned SMAI journal.

Architecture. Of course, some of the partners had a French version of project Euclid in mind while CEDRAM was under development. But this is not what was devised *in fine*: while project Euclid is an electronic edition *back-end* that will accept whatever the participating journals produce, and build the electronic offer on top of that, we conceived CEDRAM as a set of modular *plug-ins* that operate at various crucial steps in the production process, so that end-user features are uniform.

One of the main reasons for this option was that we wanted to be upwardly compatible with NUMDAM, so that the CEDRAM production could be archived by converting the metadata to NUMDAM format, which would then provide long-term access (notice that all CEDRAM journals have agreed on the moving wall model for eventual open access). Another reason was that we also wanted to provide preservation for the source files so as to have two different archiving strategies: NUMDAM for long-term accessibility to user files; and some black-box system for long-term preservation of the production system, in order to keep the sources nearest to the mathematical meaning published, rather than its representation in some ephemeral format.

So far, the existing modules that are already in use are:

1. RUCHE, a software for managing the editorial process, from paper submission and refereeing through a web-based interface to the preparation of published volumes [143];
2. cedram.cls, a LaTeX+BibTeX driven production environment [42] that automates whatever can be automated (e.g., page numbers, metadata generation)—the paper edition is a by-product of the whole process when using this system;
3. an environment that prepares all required data in the ad hoc formats for updating the electronic edition;
4. website development with a content management system that allows each journal to maintain the static pages of its website;
5. EDBM, a database, indexing, matching, linking, interactive searching software suite that takes care of the user interface to the collections (essentially the same system as in NUMDAM, which has been made more customizable).

Step 3 has been profoundly redesigned in 2007 thanks to the second generation of the CEDRAM environment. It outputs, directly from LaTeX source, XML metadata ready for generating dynamic content in XHTML+MathML, thus avoiding our legacy Latex2html-style images for mathematical expressions in titles and abstracts.

4.4 Registries, Lists, and Catalogs

We will not repeat here what has already been written elsewhere [40, 220] about the exponential growth of the mathematical literature, and the related need for adapted discovery mechanisms, going from bibliographical lists and catalogs, to systematic reviewing journals, then ending with relational databases.

The latest challenge is that there are now so many concurrent databases, each with its own set of metadata, user interface, and document coverage, that people started all over again to write lists of databases, and we are now under the pressure of maintaining either a database of databases (registry) or unified access points to the content of those databases using the greatest common denominator of their metadata set, which is very low.

We give here three examples that may only add to the whole entropy of the databases' universe.

Gallica-Math. The BNF's server Gallica has a huge amount of valuable mathematics that is somewhat hidden by a weak metadata policy. It is mostly based on a paper library catalog, which means that only physical volumes possess a record, but users deal with logical units, which might be one volume monographs, multivolume works, single contributions published inside a journal, proceedings volumes, etc.

MathDoc has built a user front-end to Gallica's resources where the available volumes are indexed at the article level. Two collections are involved so far:

1. Some important mathematicians' collected works;
2. The early years of Liouville's *Journal*, for which Gallica supplied HTML tables of contents.

Each logical item inherits an identifier and a full record so that third parties (including our mini-DML, see below) can link to them.

LiNum (*Livres Numérisés Mathématiques*). This is a consolidated database registering 2,679 freely accessible books, and 651 digitized but copyrighted, provided by Gallica (Paris), *Digital Math Books Collection* (Cornell), *Historical Math Collection* (Ann Arbor), *Mathematica* (Göttingen), *Biblioteka Wirtualna Matematyki* (Warsaw), and smaller digitization centers. It was meant to be updated on a regular basis, but only Gallica provides us with notification of its recent additions.

RBSM: The *Répertoire Bibliographique des Sciences Mathématiques*. This is a very special list of mathematical items [220] that was assembled at the turn of the twentieth century (1894–1912). It was meant as a list of articles published during the nineteenth century that were expected to be valuable to future research. Papers were grouped into files, sorted according to a specific classification scheme.

The MathDoc version, which amounts currently to half of the whole document, is a searchable database, which might be useful for various historical studies. Moreover, links to the actual articles have been added when available, possibly achieving the dream of Henri Poincaré: the RBSM will ultimately act as a gateway to the selected articles.

This work is a collaboration among Gallica, Paris (scan of the files), the *Laboratoire de Philosophie et d'Histoire des Sciences*, Nancy (structured keyboarding of the cards), and MathDoc (database, indexing, online interface).

4.5 Integrating Bits and Pieces: The Mini-DML

We have seen that the comprehensive DML is not here, and will not soon be a reality. There are many reasons, some of them obvious, others quite obscure, that prevent us from building today a navigable web of the mathematical knowledge. However, to the naïve mathematician's eye, it seems that we cannot expect more than a world where each stakeholder takes care of the collections it is, or feels, responsible for: publishers provide the best access and navigational features to their current offer, librarians and archivists populate secure and stable repositories, reviewing databases collect all relevant metadata, aggregators of any sort provide instantiated views and personalized discovery tools for the collections. We cannot believe that there will be some day one central access point to such a disorganized system; we even might be led to think that it is not the ideal situation when the actual usability of a fundamental resource depends upon a group that is too restricted (we could compare the robustness of the Internet, thanks to its web structure, with the fragility of the DNS system).

Since the announcement of the prototype for a mini-DML in 2004 [40], we have developed both the internal structure and the content so that we can consider it already a useful service—one of those foundations upon which the actual DML will be built. This can be defined as unified indexation of articles available in digital format, taking advantage of the general dissemination of OAI-PMH (open access initiative protocol for metadata harvesting) technology. A simple search interface is up and running, providing seamless access to more than 232 000 articles from 11 sources.

The mini-DML is not a competitor to the traditional reviewing databases; it is a register of the basic bibliographic metadata associated with mathematical articles in digital form, as provided by their publishers. It allows for one-stop searching over multiple collections: for instance, one can find the articles of the *Annals of Mathematics* that have been archived at JSTOR, arXiv when it was an overlay journal, and project Euclid that now incorporates also the arXiv-EMIS period; one can at the same time find Euler articles from its collected works available at Gallica and the current English translation ongoing at arXiv; in fact, it is the only place where one can search *Portugalia Mathematicæ* and obtain links to the article's full texts digitized by Portugal's national library. It also provides matching of bibliographical references, thus making it possible to add direct links from a reference to an electronic version of the text itself. This last feature is not yet public, but it is already in use at NUMDAM, and provides more than 36,000 direct article external links, which is slightly less than the number of direct NUMDAM links, so that about 21% of our references already have a digital article associated to them.

There are still methods to explore in order to make the system more robust and more useful: we will need to standardize the minimal scheme used by DML partners to share metadata (this is a work in progress where Cornell and Göttingen libraries cooperate with Math-Doc), and to document what we did so far so that others starting similar ventures can benefit from our experience rather than start from scratch. As the DML grows, and the data comes in, we will need something like self-matching in order to group duplicates (we already have quite a few duplicates, e.g., when articles appear in digitized journals as well as in collected works).

We are expecting to add many references in the short term, but there are many valuable article repositories that do not make their metadata easily retrievable. We are still in the process of trying to convince stakeholders that it benefits them to have as many ways of accessing their scientific content as possible, and that the best way to achieve this nowadays is to deliver enough of their metadata so that third parties can generate useful links to their resources, which in turn adds visibility, and thus value, to the content they host.

4.6 MathDoc Services' URLs

- CEDRAM [url:80]
- GALLICA-MATH [url:46]
- mini-DML [url:41]
- NUMDAM [url:147]
- RBSM [url:37]

The DML-CZ Project: Objectives and First Steps

M. Bartošek, M. Lhoták, J. Rákosník, P. Sojka, and M. Šárfy

Encouraged by the idea of the World Digital Mathematics Library and by digitization activities worldwide, the Czech Mathematical Society initiated the digitization project "DML-CZ: Czech Digital Mathematics Library" to ensure availability of mathematical literature that has been published throughout history in the Czech lands, in digital archival form. Five specialized groups from different institutions cooperate to carry out the project. There are first achievements, as well as some questions, that emerged in the course of the first 18 months. In particular, the Metadata Editor has been developed as a specialized software to facilitate the process of creation, revision, and validation of metadata obtained in an automated way.

1 Motivation

Several stimuli for the Czech mathematics digitization activity may be traced: The Czech Mathematical Society (CMS) has been involved in cooperation with Zentralblatt since 1996, when the Prague Editorial Unit was established under its auspices as one of the first external collaborating units. In 2002–2003 we skimmed digitization while contributing a little to the project ERAM by digitizing about 4,000 pages of the *Jahrbuch Über die Fortschritte der Mathematik*. In 2003, Bernd Wegner inspired the CMS to take part in the (unfortunately, not approved) DML-EU project coordinated by the European Mathematical Society. All this experience proved useful in 2004 when the Academy of Sciences of the Czech Republic (AS CR) launched the national research and development program Information Society. Encouraged by the idea of the World Digital Mathematics Library (WDML) [?] and by digitization activities worldwide, the CMS initiated the proposal of the digitization project "DML-CZ: Czech Digital Mathematics Library." The aim was to ensure availability of mathematical literature that has been published throughout history in the Czech lands, in digital form. The project, proposed for the period 2005–2009, has been approved [26] and has already achieved its first promising results.

2 Goals and Partners

The goals of the project are to investigate, develop, and apply techniques, methods, and tools that will allow the creation of an infrastructure and conditions for establishing the Czech Digital Mathematics Library (DML-CZ). The mathematical literature that has been published by various institutions in the territory of the Czech lands comprises several journals of international standing, a number of conference proceedings volumes, monographs, textbooks, theses, and research reports. The journals include *Mathematica Bohemica* (formerly *Časopis pro Pěstování Matematiky*), *Czechoslovak Mathematical Journal*, and *Applications of Mathematics* (formerly *Aplikace Matematiky*) published by the Mathematical Institute AS CR in Prague, *Commentationes Mathematicae Universitatis Carolinae* published by the Charles University in Prague, *Archivum Mathematicum* published by the Masaryk University in Brno, *Kybernetika* published by the Institute of Information and Automation AS CR in Prague, and a few others. Some of them have, in the meantime, been scanned in the SUB Göttingen and we plan to apply the developed procedures and tools also on this material. In view of the common history, the ongoing close cooperation, and the linguistic closeness of both nations of the former Czechoslovakia, we plan that the suitable Slovak mathematical literature will be included as well. Besides the retrodigitized material, the existing born-digital material will be included and proper arrangements will be proposed for mostly automated incorporation of the future literature produced in electronic form. The estimated extent of the relevant literature ranges from 150,000 to 200,000 pages. Upon its completion, the DML-CZ will be integrated into the WDML. The techniques and tools developed for the DML-CZ might be later used for digitization in other fields of science.

The project team consists of five closely cooperating groups from different institutions, each of them specialized in certain aspects of the digitization project. The Mathematical Institute AS CR in Prague has taken the role of the project coordinator and deals with selection, and preparation, of materials for digitization; IPR and copyright issues; and will eventually be in charge of the maintenance of the developed DML-CZ system. The Institute of Computer Science of Masaryk University in Brno is responsible for overall technical integration, development of the digital library, coordination of metadata provision, and incorporation of the DML-CZ into the WDML. The Faculty of Informatics of Masaryk University in Brno is focused on OCR processing, techniques for searching and presenting digital documents, presentation formats, and relevant technology development and testing. The Faculty of Mathematics and Physics of Charles University in Prague deals with user requirements, metadata specifications, and linkage to Zentralblatt MATH and MathSciNet. This group, together with the coordinator, represent the mathematicians' view of the project. The Library AS CR in Prague ensures digitization, OCR, and storage and presentation of the digitized content within the Academy of Science framework.

3 DML-CZ Implementation

Do not reinvent the wheel!

—Petr Sojka

We have analyzed [241] solutions used in other digitization efforts, especially those ongoing in Grenoble (NUMDAM) and Göttingen (DIEPER) and proposed the general scheme for the project, as depicted in Figure 1.

The page scanning is done in the Digitization Center of the Library AS CR.[1] With the support from the EU Solidarity Fund, the center was equipped with three high-quality book scanners and specialized software for scanned image processing. Its current production output is some 50,000 pages a month. For the DML-CZ project, two Zeutschel OS 7000 scanners (approximately 90 A4 pages per hour at 600 DPI) are used.

The test bed for the DML-CZ consists of digitized documents from the *Czechoslovak Mathematical Journal*. This journal, published since

[1] The Digitization Center was established after the devastating floods in 2002 that affected a large number of library institutions in the Czech Republic.

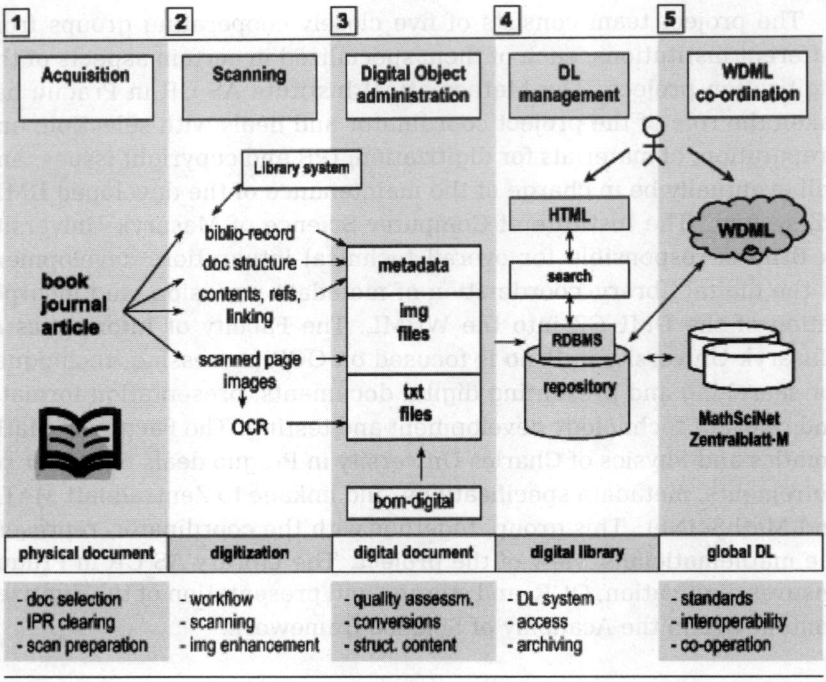

Figure 1. General scheme of the DML-CZ project.

1951, changed its layout several times, as well as its publishing policy. The first two volumes were published simultaneously in Czech, Russian, and multilingual versions. Beginning with the third volume, only one multilingual version has been published containing papers, abstracts, book reviews, and advertisements written in different languages including Czech, Slovak, Russian, English, German, French, and Italian. Individual items can begin or end in the middle of pages. A typical multilingual page can be seen in Figure 2.

Nowadays, the content is almost exclusively in English. The papers contain free-hand drawings, graphic figures, tables, and photographs. Classical typesetting was used until 1991, since then the journal has been typeset in $\TeX$; the corresponding digital files are available.

This test bed, consisting of almost 30,000 scanned pages, has been used for the development of an automated dataflow for a complete processing of mathematical documents—starting with the scanned images and finishing with the enhanced documents (articles and interlinked metadata records) in the repository.

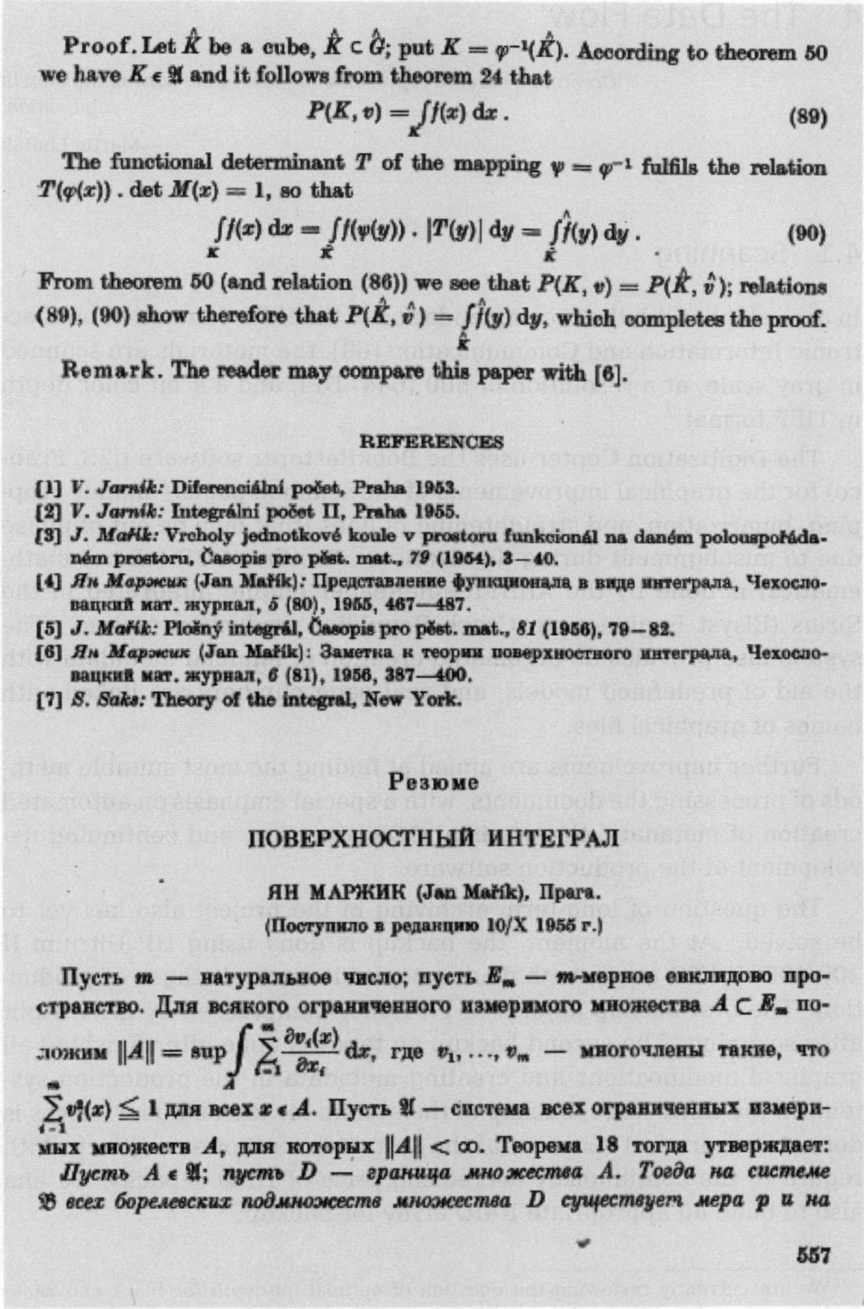

Proof. Let $\hat{K}$ be a cube, $\hat{K} \subset \hat{G}$; put $K = \varphi^{-1}(\hat{K})$. According to theorem 50 we have $K \in \mathfrak{A}$ and it follows from theorem 24 that

$$P(K, v) = \int_K f(x)\,\mathrm{d}x\,. \tag{89}$$

The functional determinant T of the mapping $\psi = \varphi^{-1}$ fulfils the relation $T(\varphi(x))\,.\,\det M(x) = 1$, so that

$$\int_K f(x)\,\mathrm{d}x = \int_{\hat{K}} f(\psi(y))\,.\,|T(y)|\,\mathrm{d}y = \int_{\hat{K}} \hat{f}(y)\,\mathrm{d}y\,. \tag{90}$$

From theorem 50 (and relation (86)) we see that $P(K, v) = P(\hat{K}, \hat{v})$; relations (89), (90) show therefore that $P(\hat{K}, \hat{v}) = \int_{\hat{K}} \hat{f}(y)\,\mathrm{d}y$, which completes the proof.

Remark. The reader may compare this paper with [6].

REFERENCES

[1] *V. Jarník:* Diferenciální počet, Praha 1953.
[2] *V. Jarník:* Integrální počet II, Praha 1955.
[3] *J. Mařík:* Vrcholy jednotkové koule v prostoru funkcionál na daném polouspořádaném prostoru, Časopis pro pěst. mat., *79* (1954), 3—40.
[4] *Ян Маржик* (Jan Mařík): Представление функционала в виде интеграла, Чехословацкий мат. журнал, *5* (80), 1955, 467—487.
[5] *J. Mařík:* Plošný integrál, Časopis pro pěst. mat., *81* (1956), 79—82.
[6] *Ян Маржик* (Jan Mařík): Заметка к теории поверхностного интеграла, Чехословацкий мат. журнал, *6* (81), 1956, 387—400.
[7] *S. Saks:* Theory of the integral, New York.

Резюме

ПОВЕРХНОСТНЫЙ ИНТЕГРАЛ

ЯН МАРЖИК (Jan Mařík), Прага.

(Поступило в редакцию 10/X 1955 г.)

Пусть m — натуральное число; пусть E_m — m-мерное евклидово пространство. Для всякого ограниченного измеримого множества $A \subset E_m$ положим $\|A\| = \sup \int_A \sum_{i=1}^{m} \frac{\partial v_i(x)}{\partial x_i}\,\mathrm{d}x$, где $v_1, \ldots, v_m$ — многочлены такие, что $\sum_{i=1}^{m} v_i^2(x) \leqq 1$ для всех $x \in A$. Пусть $\mathfrak{A}$ — система всех ограниченных измеримых множеств A, для которых $\|A\| < \infty$. Теорема 18 тогда утверждает:

Пусть $A \in \mathfrak{A}$; пусть D — граница множества A. Тогда на системе $\mathfrak{B}$ всех борелевских подмножеств множества D существует мера p и на

557

Figure 2. Typical multilingual CMJ page.

4 The Data Flow

Creating metadata represents the most time-consuming step in digitization.

—Martin Lhoták

4.1 Scanning

In accordance with the recommendation of the IMU Committee on Electronic Information and Communication [69], the materials are scanned in gray scale, at a resolution of 600 (644) DPI, and a 4 bit color depth in TIFF format.[2]

The Digitization Center uses the BookRestorer software (i2S, France) for the graphical improvements of the scanned pages—mainly cropping, binarization, and straightening of lines (that may be out of phase due to misalignment during digitization). The first OCR (all but mathematics) is done by the ABBYY FineReader engine integrated in the Sirius (Elsyst Engineering, Czech Republic) production system. The system also provides an automated creation of minimal metadata with the aid of predefined models, and real page numbers are linked with names of graphical files.

Further improvements are aimed at finding the most suitable methods of processing the documents, with a special emphasis on automated creation of metadata, the maximal OCR precision, and continuing development of the production software.

The question of long-term archiving in the project also has yet to be solved. At the moment, the backup is done using HP Ultrium II 200/400GB LTO tapes, with double copies in several stages of production. The first backup applies to TIFF files compressed by LZW, right after scanning. The second backup on tapes is done after finishing all graphical modifications and creating metadata in the production system. In the third stage, backup of the system for the end-user access is done, to ensure fast recovery of the system in case of emergency. With regard to the continuously decreasing price of HDD capacity, we aim also to build an appropriate RAID array for backup.

[2]We are currently reviewing the question of optimal bit-depth for black and white document scanning—whether or not decreasing the color depth to 1 bit in the scanning phase would have a significant negative impact on the OCR quality (see [242]). Going down to 1 bit would speed up the work flow and lower the costs of digitization.

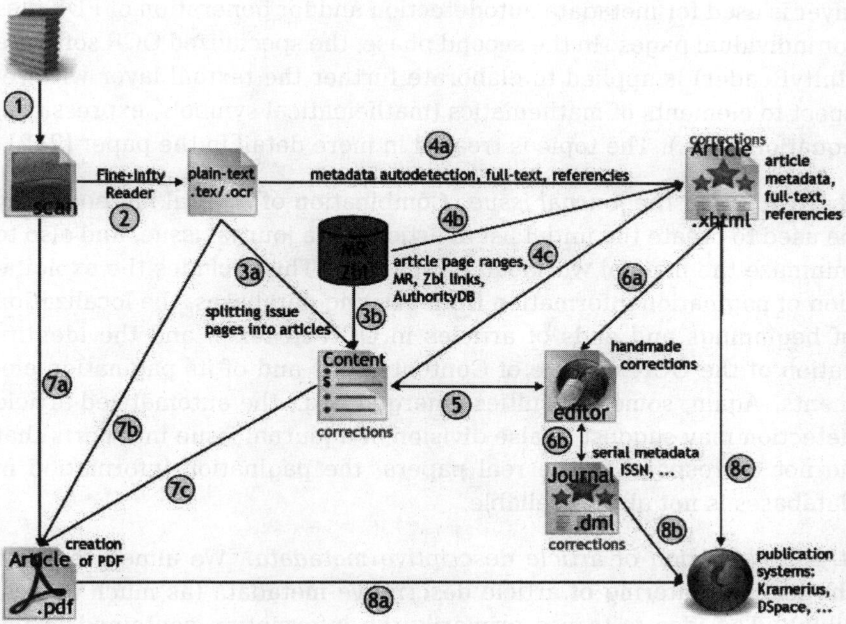

Figure 3. Automated work flow: from scanned images to articles.

4.2 From Scanned Images to Articles

Design and implementation of a fully automated environment for processing mathematical documents (journal papers at the present stage) are the tasks of the Masaryk University team. The scheme of the developed work flow as depicted in Figure 3 contains the following steps:

1. Preparation of scanned data. All the data produced in the scanning phase (page images, initial structural metadata, and page metadata) are validated with respect to their completeness and consistency, page order correctness, duplicates, etc. The data are then restructured and stored in the hierarchical Journal–Volume–Issue directory structure suitable for further processing.

2. Application of advanced OCR techniques for processing mathematical terms [252], [251]. The OCR process is realized in two phases: in the first phase, the general OCR software (FineReader) is used for the text recognition with language detection on the paragraph level (required by the multilingual nature of articles in CMJ). Resulting textual layer is used for metadata autodetection and for generation of PDF files for individual pages. In the second phase, the specialized OCR software (InftyReader) is applied to elaborate further the textual layer with respect to elements of mathematics (mathematical symbols, expressions, equations, etc.). The topic is treated in more detail in the paper [242].

3. Structuring the journal issue. Combination of several methods may be used to create the initial list of articles of a journal issue, and also to minimize the manual workload in this step. This includes the exploitation of pagination information from existing databases, the localization of beginnings and ends of articles in OCR-ed texts, and the identification of the OCRed Table of Contents page and of its pagination elements. Again, some difficulties emerged, e.g., the automatized article detection may suggest a false division of a journal issue into parts that do not correspond to the real papers, the pagination information in databases is not always reliable.

4. Autodetection of article descriptive metadata. We aimed to avoid the manual entering of article descriptive metadata (as much as possible). The idea is to use primarily the information contained in the reference databases—in particular, Zentralblatt MATH and Mathematical Reviews—to compare it and to complete it with the metadata mined from the OCR-ed text layer. Notably, in the case of older historical journal volumes, we cannot rely on the metadata from databases only and more sophisticated methods of metadata extraction must be used.

A list of references represents the important part of article metadata. An automated identification of the block of references in the OCR-ed text is a relatively easy task; its beginning can be usually detected by localizing an appropriate keyword (References, Bibliographie, Literaturverzeichnis, Littérature, Literatura, Литература, etc.) in any of the languages used in our mathematical journals. Separation of individual references and recognizing their internal structure is a much more difficult problem and further research is needed to improve the quality of the process. Once a reference item is identified and structured, a linkage to reference databases can be established.

5. Manual checking of journal issue structure. The previous steps should provide as much data as possible in an automated way. However, as we have indicated, this cannot be done in an absolutely reliable way. The manual revision and necessary corrections of data cannot be avoided and it represents a very important step in the data flow. To make it as easy as possible, the Metadata Editor has been developed.

The software provides the operator with an effective support, enabling a visual inspection and correction of the automatically generated structure of a journal issue (e.g., cancellation of badly identified articles and constituting the missing ones). Moreover, the Metadata Editor allows one to check and revise configuration of individual articles in terms of page images assignments. This includes a visual inspection of page image content, verification of the page ordering, reshuffling pages within an article and/or between articles, removing blank pages, etc.

The Metadata Editor is designed in a very user-friendly way. The operator works with the page thumbnails arranged on a grid as though they were laid on a desk, as can be seen in Figure 4.

If any correction to journal issue structure has been implemented in this step, the process returns back to step 4 (autodetection of article descriptive metadata).

6. Manual revision of article descriptive metadata. The Metadata Editor is also used to check and edit article metadata records. This step is important for the quality of the data stored in DML-CZ, not for the work flow itself. To avoid data inconsistencies, the key metadata elements (authors' names, MSC codes) are validated against appropriate authority files in the course of editing. The validated metadata record can be collated with the appropriate scanned page that is displayed in a parallel window to ease the process of visual checking and cut and paste editing; see Figure 4.

The proposed metadata structure conforms, in general, to the WDML recommendations.

Several questions emerged with respect to cataloging rules, especially in areas where unifying WDML standards are still missing—for example, romanization rules for non-Latin names and general WDML name authorities. For the purpose of the DML-CZ, a suitable combination of the transliteration tables Zbl-new and MR-new will be used.

7. Generating article PDF files. An ultimate goal is to create searchable PDF files combining image and textual layers for all articles. This is done automatically using the list of papers and corresponding page PDF files generated in previous steps. At the moment we have decided not to create DjVu files because the format is not so widely used among end users and its future is still unclear. Moreover, the declared advantage of DjVu's better compression is diminished by the fact that recent versions of PDF support JBIG2 compression.

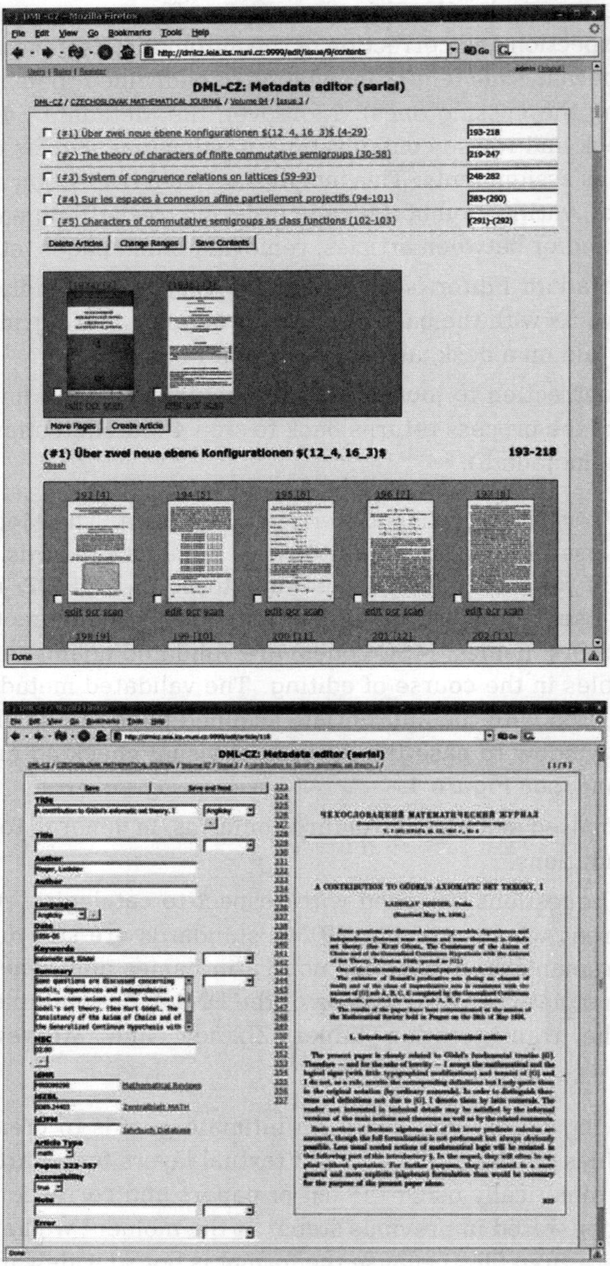

Figure 4. Metadata Editor: issue structure editing and article metadata editing.

8. Exporting papers and metadata into publication systems. The Kramerius system [166] is currently used for presenting the digitized material to end users.[3] Other digital library oriented systems or repositories, for instance the DSpace, are under consideration to host the DML-CZ as well. We shall examine also the possibility of developing a completely new presentation system specialized for the specific needs of the DML-CZ. Whatever solution is to be used, it will support the future incorporation of the DML-CZ into the WDML.

5 Conclusion and Further Steps

In addition to the first successful achievements, there are still numerous tasks to be tackled and problems to be solved in the DML-CZ project. Some of them have already been mentioned and we can list a few more:

- *IPR issues.* Even though it is possible, in general, to rely on the positive attitude of both the publishers (universities, academy institutes, the Union of Czech Mathematicians and Physicists) and authors in granting their permissions to make the digitized document freely available, there exists certain risk following from the local legislation that is not really helpful for the scientific community. We are exploring measures to minimize the limitations and to create an undepreciated digital library.
- *Further literature to be digitized.* We shall apply our experience, and tools developed in the test bed, to incorporate further journals, conference proceedings, textbooks, theses, and possibly monographs into the DML-CZ.

[3]The Kramerius system has been developed for the Czech National Library as open source software under the GNU GPL license. The Library AS CR cooperates with the Czech National Library on its enhancement. It consists of Linux, the Apache web server, the Apache Tomcat application server, and the PostgreSQL database server. XML files with metadata, together with picture files, are imported into the system. At the moment, the system supports the DjVu, PDF, JPG, and PNG file formats. The dynamic export of a few pages to one file is possible using PDF. The system supports the OAI-PMH protocol and we anticipate that the system will support other standards that will ensure better interoperability and persistent identification, e.g., OpenURL, PURL, and DOI. Both the Library AS CR and the Czech National Library use the Convera RetrievalWare engine for search within the system, which facilitates pattern search and works with Czech semantics for both metadata and full texts. To continue the development of Kramerius as freely accessible fully functional software, freely accessible searching tools are used as standard. It has already been used by several other Czech libraries.

- *Existing born-digital material.* To handle the existing born-digital material, we shall design a corresponding work flow. In particular, we plan to apply methods to enhance the files of the Czech journals digitized by the SUB Göttingen.
- *Future born-digital material.* To handle future born-digital material, we shall negotiate with the publishers and suggest arrangements in order to improve the technical quality of journals and to enable a reasonably easy incorporation of the material into the DML-CZ. The French project CEDRAM [url:80] may provide an inspiration.
- *Slovak journals and other literature.* This material should be processed as well, once we receive agreements from the corresponding publishers.
- *Citation references detection and linking.* We shall finish the tools for citation autodetection within individual papers. The aim is to enrich an article metadata record with a list of references linked to mathematical reference databases.
- *PDF/DjVu formats.* It is yet to be decided whether JBIG2 (about 20 kB per page) or CCITTFaxDecode compression (about 60 kB per page) will be used for the final presentation in the PDF format.

We are convinced that we can find solutions to all these problems and thus take advantage of the rare opportunity offered by the support of the DML-CZ project. In particular, we rely on cooperation with other digitization initiatives to tackle more complex problems, namely, OCR of mathematics, indexation and search in mathematics, classification (completion of missing MSC codes, cf. [244]), and reference linking. On the other hand, we will be happy to share our experience and developed tools. The URL for the Czech Digital Mathematics Library is [url:97].

The DML-E Digitization Project and Related Topics

The Spanish Initiative

E. Macías-Virgós

The aim of this chapter is to present the current situation and circumstances of mathematical research in Spain, some bibliographic indices, the (short) history of the Spanish digitization project DML-E, and a personal reflection about digitization matters. I hope that this report will help teams facing analogous problems in other countries. At the end of the report, I provide a rather complete bibliography (mostly in Spanish) that I hope will be of interest to our colleagues and librarians in Latin America.

1 Mathematical Research in Spain

1.1 Some Figures

The situation of mathematical research in Spain can be summarized as follows. There are 350 research groups, mainly working at universities, which receive an average of 50,000 euros every three years. The mean size of groups is five to six researchers. This investment is made available under the so-called National R+D+i Program [181]; in many cases, there are supplementary funds from the governments of local autonomous regions or from universities.

Several recent reports all underline the remarkable evolution of mathematics in Spain [9, 38]. Mathematical research has rapidly increased in quantity as well as in quality and impact. As a matter of fact, the Spanish contribution to the worldwide published papers in mathematics (measured by papers in ISI journals having a Spanish author)

93–97	94–98	95–99	96–00	97–01	98–02	99–03	00–04
3.46	3.66	3.88	4.18	4.42	4.53	4.65	4.82
-17	-14	-15	-16	-13	-13	-6	-3

Table 1. Spanish percentage production in mathematics and relative impact factor in comparison with the world mean; years 1993–2004. Source: Web of Knowledge Thomson-ISI. Taken from [38].

1990	1991	1992	1993	1994	1995	1996	1997	1998	1999
1.7	2.0	2.2	2.2	2.1	2.3	2.4	2.7	2.8	3.2
1.7	1.7	2.1	2.6	2.5	2.9	3.0	3.3	3.6	3.9

Table 2. Percentage comparison Spain/world production in mathematics; years 1990–1999. Source: MathSciNet (first row) and ISI (second row). Taken from [9].

rose from 0.3% in 1980 to the current 5.0%. As a reference, the overall Spanish percentage share for all scientific fields is 3.18%. Spain now occupies the ninth world position in terms of contribution to mathematics, which more or less corresponds to our economic status.

According to [9], in the period 1996–2001, there were 5,600 different Spanish authors of mathematical papers (60% of them published just one paper). The median number of authors was two; and 40% of papers were written in international collaboration (mainly with European countries and North America, and increasingly with Latin America).

On the other hand [38], for the period 1994–2004, the total number of papers (in ISI journals) was 6,220. Approximately 20% of them were in the first quartile of journals ranked by impact factor (IF, number of cites per document) in JCR Web of Science. The mean IF was slightly below the world mean (2.28 versus 2.51).

This rather impressive development has received international recognition, e.g., with the organization of the 3rd European Congress of Mathematics (Barcelona, 2000), and the choice of Spain by IMU as the venue of its 15th General Assembly (Santiago de Compostela) and the International Congress of Mathematicians (ICM Madrid, 2006).

1.2 Consolider Mathematica

However, several weaknesses have been detected in the Spanish mathematical research, including the lack of an adequate contact with industry; the small presence of computational infrastructures to support research (with some brilliant exceptions [59]); insufficient long-term planning; and a low leading role in many fields.

Hence, a new program from the Ministry of Education and Science has been approved [70] under the name of Consolider Ingenio 2010. The mathematics branch of this program, called Mathematica (not a very original name!), will offer the mathematical community a supplement of 7.5 million euros for the next five years (renewable for five more years).

The project aims to develop strategic measures in order to provide cohesion to the activity of Spanish mathematicians; to increase the weight of mathematics in the Spanish system of science, technology, and industry; and to improve our situation on the international stage.

Mathematica is structured in five nodes[1] with a board of directors formed by fourteen leading Spanish mathematicians.

The main proposed measures include semesters devoted to specific subjects, internationally recognized doctorate courses, and fellowships for post-docs and researchers (from Spain and abroad). The tools at our disposal will be of three types:

1. Platforms, named according to their activities: Future (for identifying emergent research areas of strategic interest), Consulting (for applicability of results in a nonacademic context), Computing (a network infrastructure of computational tools and software repositories), Edu (for Mathematical Education), and Web (see below);
2. Thematic Actions, such as an International Graduate School or several Programs of Intensive Research;
3. Cross-sectional Services, e.g., to support meetings and encounters, with a recently inaugurated facility located in Cantabria.

1.3 The Platform Mathematica Web

In order to increase the accessibility of the results of mathematical research, a specific platform will be created. It will present the activities and services of the project Mathematica, as well as the corresponding links to services, research programs, thematic networks, and platforms. This "virtual house" of Mathematica will also give access to such tools of general interest as databases, specific software libraries, digi-

[1] Centre de Recerca Matemática (CRM Barcelona), Institute of Mathematical Sciences (ICMM Madrid), Institute of Mathematics (IMUB Barcelona), Centre of Supercomputation of Galicia (CESGA Santiago de Compostela), and Centre for International Mathematics Meetings (CIEM Cantabria).

tized publications, preprint servers, MathNet webpages, and research journals. Also, divulgation, diffusion, and prospective will be taken into account for public awareness of the mathematical activity.

2 Access to Scholarly Literature

2.1 Open Access

Dissemination of knowledge is not complete if the information is not readily available. In March 2006, the European Commission issued an important study on the economic and technical evolution of the scholarly publication system and on the scientific publication markets in Europe. The EC recommendations include "to guarantee public access to publicly-funded research, shortly after publication" [248].

According to [182] (see also [25]), Open Access in Spain is still an emerging movement. Currently, the 180 registered signatories of the Berlin Declaration include 18 Spanish universities or research institutions. ROAR (Registry of Open Access Repositories [216]) and OpenDOAR (Directory of Open Access Repositories [200]) have in their records 17 and 11 Spanish open access repositories, respectively, which represent less than two percent of the whole institutional repositories registered in those directories.

However, in recent years more and more initiatives related either to institutional open access repositories or open/free journals have been reported. There is an intense debate among librarians about the impact and visibility of scientific publications and the restrictions imposed by monopolistic publishers. Last summer, a course [261] about Open Access was organized by FECYT (Fundación Española para la Ciencia y la Tecnología [119]). Very recently, the symposium Libraries and Digital Objects [34] which was organized by CSIC, brought together many international experts.

Open archives serve to store papers, preprints, conference proceedings, reports, and other research documents; and include metadata that are accessible through an interface like OAI-PMH (Protocol for Metadata Harvesting). Most current repositories started as repositories for doctoral theses, but some of them have evolved to institutional repositories (IR), which serve to collect the scientific production of an institution in such a way that it remains distributed in multiple archives but with a common interconnecting tool, ensuring longterm preservation of academic work.

Open/free access journals were formerly printed journals that are currently free online [86]. In most cases they are published by public institutions or learned societies. Examples of open access journals directories or digital libraries are Scielo Spain (Scientific Electronic Library on-line [233]); Dialnet [81]; and Tecnociencia e-revistas (see below). There are other projects in several universities.

The Ministry of Culture has a directory of digitization projects and OAI harvesters [179] with basic information about the projects and initiatives of digitization existing in Spain.

2.2 Tecnociencia

Tecnociencia [254] is a website about science and technology in Spain, which depends on CINDOC-CSIC (see Section 3.3). It was created after a long discussion about the need of good scientific journals in the Latin American world, and it aims to provide unified access to electronic journals, under the auspices of FECYT. Tecnociencia includes news, research projects, and an OAI server called e-revistas offering e-journals.

2.3 ReviCien

Currently, almost 30 Spanish scientific journals are included in the Thomson-ISI Science Journal Citation Index. Four of them are from mathematics.

Most of these journals share a common online platform for promotion and diffusion, which is called ReviCien (Revistas Científicas, meaning Scientific Journals [211]). It offers links to 35 Spanish scientific journals from agricultural sciences, earth sciences, environmental sciences, life sciences, chemistry, engineering and technology, materials science, computer science, mathematics (12 journals), medicine, neuroscience, pharmacology and toxicology, physics, and astronomy.

While full texts of papers remain in the website of each journal, users of ReviCien have access to the abstracts and basic information such as publisher, editor, ISSN, scope of the journal, cover, and table of contents of current or back issues, as well as quick/advanced search by author/title/keywords.

Subscribers receive e-alerts with tables of contents and announcements, and journalists have a dedicated Press Room. Finally, journal editors can access a restricted area inside the server.

3 Organizations Involved in DML-E

3.1 The Spanish IMU Committee CEMAT

In order to put some spine into the Spanish mathematical community, a very important role is reserved for the Spanish Mathematics Committee (CEMAT, meaning Comité Español de Matemáticas [57]), which coordinates the activities of international scope in Spain related to the International Mathematical Union (IMU). It also advises the Ministry of Education and Science of Spain about IMU recommendations on education and research in mathematics, and ensures coordination with ICSU (International Council for Science). CEMAT publishes a webpage [57] and a regular newsletter.

Spain entered IMU in 1952 (group II), passed to group III in 1986, and entered group IV in 2004. The current CEMAT was founded in 2004, as a continuation and extension of the former Spanish IMU Committee. Today, the seven main Spanish mathematical societies[2] are represented in CEMAT. There are also representatives of the Ministry of Education and Science.

CEMAT has an Executive Committee, a General Council, and four Commissions, which are counterparts of the analogous IMU Committees, which are as follows:

- Commission on Development and Cooperation;
- Commission on Education;
- Commission on History;
- Commission on Electronic Information and Communication.

3.2 The Spanish CEIC

The purpose of the Commission on Electronic Information and Communication[3] is to advise the Spanish Committee of Mathematics on all subjects related to electronic communication, and to represent Spain in the activities of the Committee on Electronic Information and Communication (CEIC) of the IMU.

[2]RSME (Royal Spanish Mathematical Society), SCM (Catalan Society for Mathematics), SEMA (Spanish Society for Applied Mathematics), SEIO (Society for Statistics and Operations Research), FESPM (Spanish Federation of Associations of Mathematics Teachers), SEIEM (Spanish Society for Research in Mathematics Education), SEHCYT (Spanish Society for History of Science and Technology).

[3]Its present composition is as follows: Jaume Amorós (Secretary, Universitat Politécnica de Barcelona), Manuel González Villa (Universidad Complutense de Madrid), Rafael de la Llave (University of Texas at Austin), Juan Luis Varona (Universidad de La Rioja), and Enrique Macías (Chairman, Universidade de Santiago de Compostela).

The Commission maintains a webpage with links to interesting material such as digitized journals, literature related to electronic scientific edition, and the Spanish TEX users' group (CervanTEX [58]).

3.3 CINDOC

The Spanish Center for Scientific Information and Documentation (Centro de Información y Documentación Científica [64]) aims to provide adequate support and scientific information to Spanish users in all areas of knowledge, to collect the Spanish scientific output, and to make it widely available.

CINDOC also undertakes research projects in the field of scientific documentation; carries out bibliographic studies; develops systems, methods, tools, and techniques for information processing, storage, retrieval, and dissemination of information; and organizes training courses in order to encourage the use of the information technologies.

CINDOC depends on CSIC (Consejo Superior de Investigaciones Científicas), the Spanish National Research Council [72]. Its personnel is comprised of about 100 librarians and technicians.

3.4 A Brief History

The Information and Documentation Center (CID) was created in 1953 as a bibliographic inquiry service that began the publication of an Index to Scientific and Technical Journals, later replaced by the Abstracts of Scientific and Technical Papers [64].

In the 1970s a new body was created, with the name of National Centre for Scientific Information and Documentation (CENIDOC). CENIDOC was conceived as a coordinating body, operating through three sectoral institutes: ICYT (Institute for Information and Documentation in Science and Technology), ISOC (Institute for Information and Documentation in Social Sciences and Humanities), and IBIM (Institute for Information and Documentation in Biomedicine).

In 1976, ISOC began to publish both the Spanish Index on Humanities and the Spanish Index on Social Sciences, while ICYT began the publication of the Spanish Index on Science and Technology in 1979. All three indexes covered the bibliographic references for all papers published in Spanish scientific journals. Since 1989, these databases have been available on-line. In 1990, the CSIC databases were the first bibliographic information product ever released on CD-ROM in Spain.

Back in 1992, ICYT and ISOC merged to form the new Scientific Information and Documentation Centre (CINDOC).

4 The Digitization Project

DML-E is the Spanish counterpart of the WDML project of digitization of scholarly literature in mathematics supported by IMU. Since 2000, we have participated in several meetings sponsored by the EMS and were actively involved in the preparation of DML-EU proposals. Finally, it was clear that the European Commission did not intend to cover digitization fees, but only coordination tasks, definition of standards, and long-term archiving. Hence national funds were necessary to support the physical process of digitization, metadata capture, server access, and linking to databases.

In 2005, the Ministry of Education and Science agreed to fund a first digitization project in mathematics by means of a so-called MEC Special Action 2005-2007. The objective is to retrodigitize all mathematical research journals published in Spain, starting from 1980 (sometimes 1940, depending on the scientific interest), which means about 100,000 pages. The scientific responsibility relies on CEMAT while the technical part is being developed by CINDOC (see Section 3.3). There is a representative of the editors of the involved journals.

4.1 Standards

Thirteen research journals with a recognized scientific quality in mathematics participate in DML-E. The project is digitizing about 50,000 pages per year and by mid-2007 we expect to have a dedicated web portal, with Open Access metadata and summaries, stable URL addresses, and pointers to MathSciNet, Zentralblatt, and catalogs from universities.

The standards follow WDML and CEIC recommendations. Resolution is 600 DPI, with software correction of text orientation and margins. Each page is archived in a tiff bitonal lossless compression file of the type `journal/ year/ volume/ number/ paper/ page.tif`. A PDF file for each paper is composed from the TIFF files of its pages.

4.2 Metadata

CINDOC has developed a tool for capturing data and a web interface with a search engine. The structure of the metadata is as follows: title (original language, English, Spanish); author; journal; organization(s); keywords (English, Spanish); abstract (English, French, Spanish); UNESCO classification; publication year; collation; type of document (book, article, report); archive.

English	Spanish	French	Italian	German	Catalan
79.2	12.9	5.9	1.4	0.4	0.1

Table 3. Percentage of original languages in digitized papers in DML-E.

Most papers are written in English (see Table 3).

4.3 Journals

The following is a comprehensive list of mathematics research journals published in Spain:

- *Applied General Topology,* Universidad Politécnica de Valencia;
- *Collectanea Mathematica,* Universitat de Barcelona (UB);
- *Extracta Mathematicae,* Universidad de Extremadura;
- *Matemàtiques,* Universitat de València;
- *Mathware and Soft Computing,* Universidad de Granada and Universitat Politécnica de Catalunya (UPC);
- *Publicacións Matemàtiques,* Universitat Autónoma de Barcelona (UAB);
- *Qualitative Theory of Dynamical Systems,* Universitat de Lleida;
- *RACSAM Revista de la Real Academia de Ciencias Exactas, Físicas y Naturales, Serie A: Matemáticas;*
- *Revista Estadística Española,* Instituto Nacional de Estadística (INE);
- *Revista Matemática Complutense,* Universidad Complutense de Madrid; old series: *Revista Matemática de la UCM;*
- *Revista Matemática Iberoamericana,* CSIC-RSME; old series: *Revista Matemática Hispano-Americana;*
- *SORT: Statistics and Operations Research Transactions,* Institut d'Estadística de Catalunya; old series: *Qüestiió: Quaderns d'Estadística i Investigació Operativa;*
- *TEST,* Sociedad de Estadística e Investigación de Operaciones (SEIO); old series: *Trabajos de Estadística;*
- *TOP,* Sociedad de Estadística e Investigación de Operaciones (SEIO); old series: *Trabajos de Investigación Operativa;*
- *Disertaciones Matemáticas,* Departamento de Matemáticas Fundamentales, Universidad Nacional de Educación a Distancia UNED.

More information about these journals appears in my Stockholm report [174] and on their webpages. I have not included journals from mathematical societies, e.g., *La Gaceta de la RSME, Butlletí de la SCM.*

5 Previous Reports

The following references are pertinent for understanding the evolution of the digitization initiatives for mathematics in Spain. First, my report to the Royal Spanish Mathematical Society (RSME) [173] was largely inspired by John Ewing's report [102], but includes my own experience when preparing the EMS applications for EU funds.

In the same issue of *La Gaceta* there is a very good paper from Rafael de la Llave [76] explaining several key ideas about scientific publication, universal access, the role of librarians, and copyright issues. Rafael de la Llave makes very clever comments on distribution and archiving of scientific information, editorial policies, and evaluation of journals' quality.

Another interesting report is the CINDOC publication [212] about the state of the art in electronic publishing.

Finally, there is an excellent book by José Luis González Quirós and Karim Gherab [124] about the so-called Universal Digital Library. The authors (a philosopher and a theoretical physicist) give a very detailed account on the current status of electronic publishing and related topics, but mainly they explore with great insight the profound changes that digitized repositories will introduce in our experience as scientists and readers.

Digital Libraries and the Rebirth of Printed Journals

J. Borbinha

The emergence of the Internet, and of new ways to produce new content and services associated with the dissemination of cultural and scientific information, has been promoting a new paradigm: the "digital library." Here, we will describe a generic model of the digital library, according to the most common trends. This model will be used as a reference point to discuss relevant technical and business issues related to digital publishing, with special attention to the issues of metadata and interoperability. The discussion will be followed by the practical example of the republishing of *Portugaliæ Mathematica*, a journal of the Portuguese Mathematical Society (SPM) [url:165]. The journal is currently published in both printed and digital media, but the first 50 volumes were published only in print. These volumes were digitized in a partnership between the SPM and the National Library of Portugal (BNP), and integrated into the National Digital Library (BND) [url:8]. The metadata is available through a server compliant with the Open Archives Initiative–Protocol for Metadata Harvesting (OAI-PMH) [url:150], and also through other simple solutions coded in XML, making the papers in these volumes searchable in emerging services such as Mini-DML [url:41], the European Library (TEL) [url:173], and Google Scholar. Searches are also possible trough the Z39.50 server of the BND. Finally, the papers are also linked to their descriptions in both MathSciNet [url:69] and Zentralblatt MATH [url:195].

1 Digital Libraries

There are many possible definitions for a "digital library." Our purpose here is not to discuss or propose any in particular, but instead raise the awareness for its fundamental business processes.

Figure 1 illustrates an evolutionary view of this scope, taken from the perspective of the traditional library. In a continuum of events and revolutions, we can point to the computer and the Internet as the most recent relevant factors in the technological evolution of the "library." The computer (initially the mainframe or the minicomputer) had a first impact on the cataloging of, and on the definition of the actual standards for, bibliographic information (resulting in the actual concept of descriptive metadata). Following these developments, data communication networks made possible remote access to the catalog and other related services. The emergence of the personal computer and the CD-ROM promoted the digitized library, also providing access to its contents. Finally, the Internet and the World Wide Web have now brought us the ultimate vision of the "virtual library." During this evolution, the "digital library" has been a fluctuating concept, with no unique definition. The perception of it depends too much upon the moment and the perspective. In this sense, and for the purpose of our discussion here, we propose this simple definition: a controlled group of services, maintained by an identified entity, making possible the storage, registration, preservation, and discovery of, and access to, digital information re-

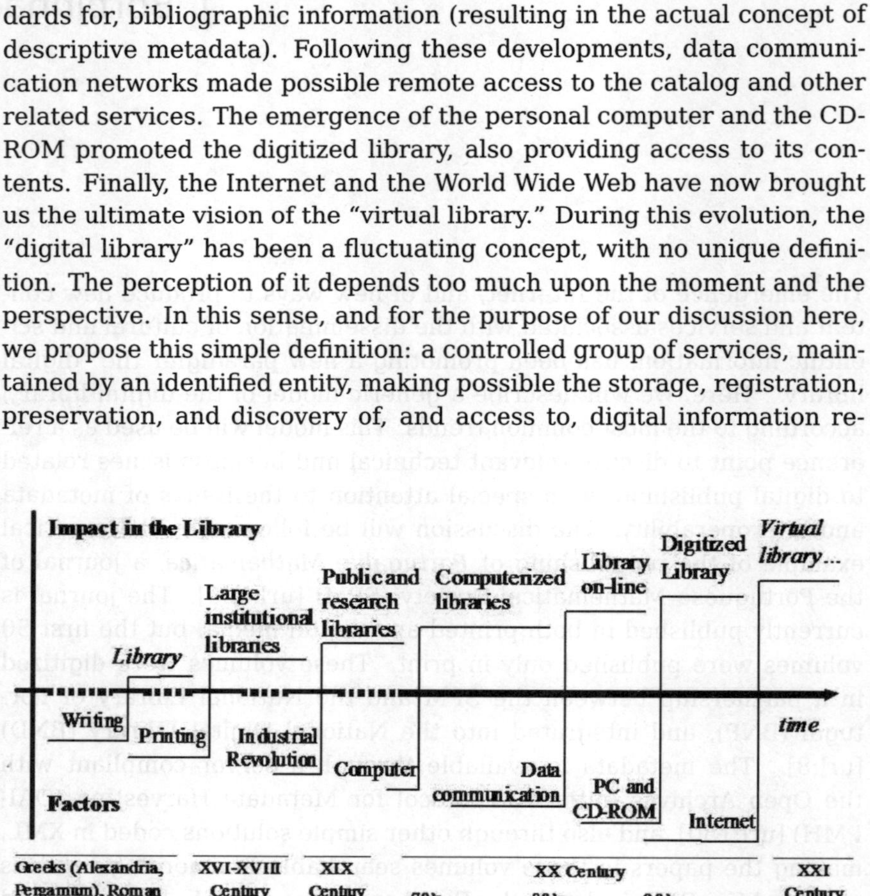

Figure 1. Libraries and technology throughout history.

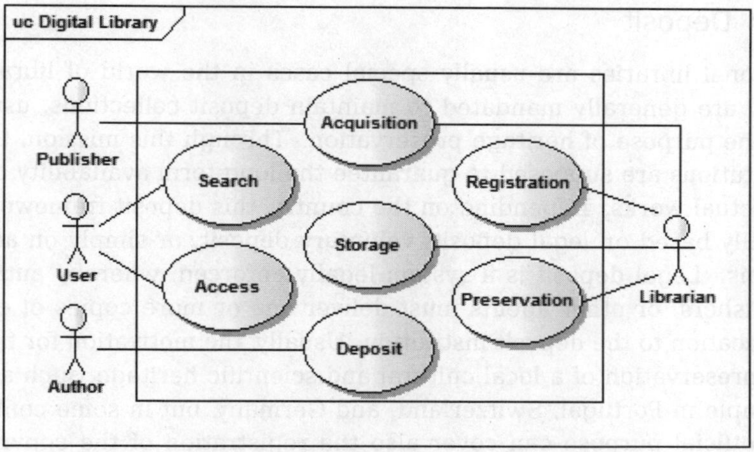

Figure 2. The main uses of a digital library.

sources. This definition can be illustrated by the diagram in Figure 2, from which we can derive the main related business processes.

1.1 Acquisition

In a digital environment it is possible for a library to acquire one digital copy of a resource, and make it easily accessible to multiple users. This is usually true for technically simple materials, such as e-books. However, for more complex materials, technical complexities of the resource or specific requirements for its local installation and access might be a serious issue (that can be the case of simulators, for example, with specific requirements for file systems, database management systems, viewers, etc.). A model that is becoming popular to minimize these problems is the licensing of remote access to the digital resources. In this case the libraries have no need to acquire physically the resource items, but they acquire the right for their users to access them remotely. This is a model with success in library's consortia, where groups of libraries can manage to impose strong conditions on the big publishers. However, this model raises some concerns in the long term: what happens to the investment made in the past by the library if it decides not to continue the subscription or if the publisher ceases to offer the service? In answer to this, some publishers still provide copies of the resources to the library, but it is not certain that such will be technically compatible in the long term (the problem of digital preservation). This is a mission for the deposit libraries!

1.2 Deposit

National libraries are usually special cases in the world of libraries. They are generally mandated to maintain deposit collections, usually for the purpose of heritage preservation. Through this mission, these institutions are supposed to guarantee the long-term availability of intellectual works. Depending on the country, this deposit framework is usually based on legal deposit, voluntary deposit, or simply on acquisitions. Legal deposit is a system legally enforced, whereby authors, publishers, or other agents must deliver one or more copies of every publication to the deposit institution. Usually, the motivation for this is the preservation of a local cultural and scientific heritage, such as for example in Portugal, Switzerland, and Germany, but in some contexts the official purpose can cover also the registration of the copyright. In these cases, the deposit institutions receive, register, and preserve those materials as proof, to guarantee to the authors the recognition of their intellectual ownership. This is the case e.g., in France and in the United States (where materials are deposited in the Library of Congress). Voluntary deposit is a system usually based on agreements between the deposit institution and the publishers or authors, under which those agents deliver one or more copies of each publication for preservation. This is, for example, the case in the Netherlands. Finally, in the acquisitions system, the deposit institutions have to take the intiative themselves to identify, select, and acquire the relevant publications, according to their mission and strategy (this occurs usually in auctions for rare books, manuscripts, or other special materials, where national laws usually give to the national libraries a right of preference for these acquisitions).

1.3 Registration

The process of registration comprises the collection or production of metadata to describe a publication and its items. This includes the traditional tasks of cataloging and indexing, and possibly new tasks identified in the future that are specific to the digital paradigm. Some tasks can be automated, reusing embedded information provided with the publications or using special software applications for automatic analysis of the materials. It is, however, necessary to address now new requirements, such as issues related to the installation and uninstallation of the digital publications (especially for those requiring executable software), the requirements for access (decryption keys or passwords, metadata for legal terms and conditions of use, etc.), and preservation (file formats, logical structures of hypermedia publications, etc.). The metadata related to these requirements is often called *management metadata* and *preservation metadata*.

1.4 Search

The purpose of the search process is to support the discovery of the existence of the works and to learn about their locations. The traditional tool used for this purpose has been the catalog, whose key conceptual roots can be traced back to the ancient Library of Alexandria and the work of its most well-known curator, Callimachus. The Online Public Access Catalog (OPAC) has been the actual tool to support this task in the era of the "networked library." It is even now possible to search simultaneously among multiple catalogs, through technological solutions such as the ISO 23950 protocol (more commonly known as Z39.50 [url:131], according to its coding given by the NISO [url:145]). Meanwhile, the Internet brought new functional opportunities for interoperability among different library management systems, augmenting the potential of this process. The OPAC is, traditionally, a "pull" service. Having it managed by computers, the libraries developed new related "push" services, such as the services commonly known as "selective dissemination of information" (by keeping a record of profiles of their users, the libraries can now use that information for focused dissemination and announcement of new acquisitions or licenses).

1.5 Access

Traditionally, the acquisition of a publication is the set of actions of assuming the ownership of at least one item of that resource. The physical ownership of the items is therefore an important requirement for a good quality of service. As an alternative, it is possible to use interlibrary loan services and request a remote reproduction, which makes it possible for a user's library to access information resources from other libraries. But those services are usually expensive and slow. Access to resources in a digital paradigm can be made easier now, since there is no need anymore to get them from the shelves. But this facilitated scenario raises the need to prevent abuse in the usage of licensed material.

1.6 Storage

The immediate purpose of storage is to make it possible for the library to provide access to the resources it acquires. In the case of licensing, this is not a necessary process. However, if the library has to hold the contents, that can become a fundamental issue. This is the case for a deposit library, for example.

1.7 Preservation

We can understand the problem of digital preservation from three main perspectives: physical preservation, logical preservation, and intellectual preservation. The entire problem starts with physical preservation. Until recently, the offline media for distribution was associated mainly with discs—CD-ROM or DVD—while the designated media for long-term storage was magnetic tape. These media have been shown to be quite fragile, requiring constant refreshment. Even the DVD has not yet proved its reliability. Also, the fact that it is an offline technology always presents a limitation to the requirement of having everything accessible at any moment. The problem of logical preservation is associated with the need to assure format conversions when original formats become obsolete or too expensive to maintain. Besides its complexity, this carries other potential problems, namely, the definition of the legal status of the new versions of the converted objects. This is a potential source of conflict with the rights' owners, especially in the entertainment industry. Finally, intellectual preservation is the problem of assuring that an information object can be exploited intellectually in the same conditions under which it was initially conceived. Format conversion, for example, can imply changes in the layout or presentation of, or the forms of interaction with, the material, and thus may imply a loss of its original intellectual content that might be declared unacceptable by the original author.

2 Information Resources

In simple terms, we can consider that a digital library has to deal with two classes of information resources: *digitized objects* and *digital born resources*.

2.1 Digitized Objects

Creating digitized copies of printed manuscripts is one of the most interesting areas of actual investment for libraries with rich collections (as it is in the case of the deposit libraries, for example). As a result, multiple digitization projects are bringing general access to a large range of information resources, from usual objects such as newspapers, journals, and printed books, to invaluable and so far inaccessible treasures such as rare drawings, incunabula, manuscripts, and maps. When created in high quality, images from these digitization efforts are usually 24 bits in color depth with resolutions from 300 to 600 dots per inch (DPI). These high-quality master images are appropriate for reproduction and local use, but are too "heavy" for usual access by the Internet. For that purpose, copies in lower resolution or lower color depth are usually derived, coded in PNG, JPEG, or GIF formats. When the objects are rich in printed text, they can also be submitted to automatic optical character recognition (OCR). This is an effective and cheap way to produce text for automatic indexing. However, for works of high relevance, this text usually needs to be corrected by expensive human intervention in order to be considered as alternative transcribed copies. As a result of all these processes, we have in the end, for each digitized work, a potential list of multiple copies. Also, these copies can range from a very simple object (just one image) to very complex objects (for example, in the case of books made of multiple chapters or parts, and digitized initially in hundreds or even thousands of image files). To manage this complexity, we therefore need to record also the logical structure of these works, which requires structural metadata.

2.2 Digital Born Objects

Besides these digitized objects, we also have nowadays new original and digital born objects. These can comprise online journals, newspapers, or e-books, published in a wide range of models, or innovative websites representing new genres of resources, yet to be named. The licensing and business models for these objects, and also their requirements for storage and preservation in digital libraries, are representing new major challenges for everyone involved.

3 Metadata

Metadata was a term coined by computer scientists and engineers from the database world to refer to structured information that describes database schema (e.g., the way the data is organized in a database). This is not how the term has been used in digital libraries and by the Internet community in general, and it is very important to be aware of that. The Internet community took the term after the emergence of the World Wide Web, and now the most common usage for it is really in the perspective of "data about data" (which once stored in a database would mean, from the previous perspective, just the data inside a database, while metadata would be the information needed to describe the organization of that database). Moreover, we should prefer for metadata the definition of "structured information about other information or resources." However, even with this definition, there are still a few common misunderstandings regarding this term. For example, it is important to stress that metadata refers to information coded according to a specific schema, and not the technology that handles it or the conceptual spaces that control the values of the information elements. In this sense, terms such as MarcXchange [url:75] and Dublin Core Metadata Element Set (DCMES) [url:102] are not metadata, but metadata schema, e.g., definitions of how to express metadata as structured information about other information or resources. In the same sense, XML in itself is just a generic language, and not metadata or even a metadata schema. With XML we can define an application context by using a Document Type Definition (DTD) or the more powerful XML Schema Language. In addition, authoritative spaces, such as indexing languages and classification systems, are also not metadata in themselves, but values or rules for finding the right values to assign to metadata elements! In conclusion, a metadata record is an information object where we can register relevant information about other objects, coded in a specific language (XML, etc.); according to a specific format or schema (UNIMARC [url:177], MARC21 [url:128], Dublin Core [url:101], etc.) and specific rules (indexing or classification languages, etc.).

3.1 Metadata and Processes in the Digital Library

In a digital library framework we can find multiple kinds of metadata, which are usually created during the registration process. Descriptive metadata, a more generic name for the traditional bibliographic metadata, relates to the description and identification of the resources, such as titles, authors, indexing terms, classification codes, abstracts, etc. Administrative metadata relates to information about the management of the resource, such as information about acquisition process and costs, and terms and conditions for reproduction and access (addresses, passwords, etc.). Preservation metadata can cover technical or management information and requirements for long-term preservation of the resources. Technical and structural metadata is about information and requirements for manipulating the resource (systems and tools to copy and explore it, for example). Finally, we can even have metadata for the administration of metadata (information about other metadata, e.g., data about creation, origin, authenticity). The bibliographic description of resources is a common issue in traditional libraries and archives, where, respectively, the MAchine Readable Catalog (MARC) family of schemas and the Encoded Archival Description (EAD) [url:127] schema are widely used. The MARC standards represent a family of very rich metadata standards traditionally used in libraries. The two most common formats in this family are MARC21 and UNIMARC . These formats are intended to be carrier formats for exchange purposes. They do not stipulate the form, content, or record structure of the data within individual information systems; they provide recommendations only on the form and content of data for when it is to be exchanged. These formats cover a wide range of metadata requirements, but they have been used especially for very rich descriptive metadata. The MARC formats have been seen as too complex and expensive for most of the digital libraries, which motivated the development of more pragmatic alternatives, such as the DCMES and Dublin Core. Even if not totally suitable for description, Dublin Core is, in fact, a very effective solution to support the process of search. However, the high value of the millions of records already created worldwide based on the MARC formats cannot be ignored. They represent a huge investment of a large community of very highly skilled professionals, and in fact they have been also used as the foundation of new digital library services.

3.2 Structural Metadata

Books have very important tangible properties, such as physical binding and structure. That information must be registered in the digitization processes. Page order, chapters and sections, and provenance and format of the original book, among other data, must be recorded alongside the new virtual book, otherwise all that information will be missing to anyone observing or studying the digital object. Part of that information is usually registered in descriptive metadata, but new requirements imply the use of structural metadata. The Metadata Encoding and Transmission Standard (METS) [url:129] is the most common format used nowadays in digital libraries for this purpose. The METS schema provides a flexible mechanism for encoding structural metadata for most of the common digital library objects. It can register or refer essential descriptive and administrative metadata, including links between these various forms of metadata. METS can therefore provide a useful standard for the exchange of digital library objects between repositories. Within the reference model of Open Archival Information System (OAIS), and depending on its usage, a METS document can be used in a deposit process as a Submission Information Package (SIP), in storage and preservation as an Archival Information Package (AIP), or in an access process as a Dissemination Information Package (DIP).

4 Interoperability

The vision of the digital library can be seen as part of the overall vision of the World Wide Web, and more specifically, of the Semantic Web. The key word in this context is *interoperability*. Interoperability means the ability to exchange information between different systems, which comprises not only the need for common protocols but also the need for common metadata structures (or at least for commonly understandable metadata structures). In an interoperable scenario, the users expect not only to reach the library from anywhere, but also to reach anything. This means that users might expect to be able to use common services to search at the same time, and transparently, a library and a film archive, and in the following moments to access books and movies created by and about, for example, Federico Fellini. In order to be able to offer services of this kind, the digital library needs to be designed as an aggregation of heterogeneous services. That requires cooperation among generic and specialized libraries and archives, museums, and other classes of organizations and actors. Once again, the ability to automate this interoperability is crucial for its cost and technical effectiveness, bringing requirements for new classes of interfaces and metadata. That interoperability has been made possible until now by protocols such as Z39.50 and SRU/SRW [url:130], for remote real-time search on OPAC systems; while to transfer metadata, the most successful solution has been the OAI-PMH protocol.

5 The Case of *Portugaliæ Mathematica*

We will illustrate now, with a specific case, how some of the issues discussed above can be applied. *Portugaliæ Mathematica* is a journal published by the SPM. It is published in digital format by SPM, in PDF format, starting with volume 61 (from 2004). Volumes 51 to 60 (1994 to 2003) are also available electronically and for free, from EMIS (the European Mathematical Information Service) in Postscript and PDF, as a file for each paper. The first 50 volumes of the journal, from 1937 to 1993, existed only in paper format. In 2004, the BNP and the SPM carried out a project to digitize these volumes and publish them in the BND, with the technological support of INESC-ID. The original printed volumes existed in collections at BNP and at the SPM, but none of the collections was totally complete. For the volumes for which it was possible to find more than one copy, one of the copies was dismembered, so each page could be digitized by an automatic feed scanner. Concerning the unique volumes, they were carefully digitized, keeping their original bindings. This process involved 15,876 pages, representing 192 issues with 1,347 different papers from 1,100 different authors. The original images were all created in TIFF format, in 300 and 150 dpi (for preservation), with derived copies created in other formats for access. All these images were structured in METS files, according to the requirements of the BND. A generic OCR process was also applied to every page. This resulted, for each paper, in a preservation copy, in

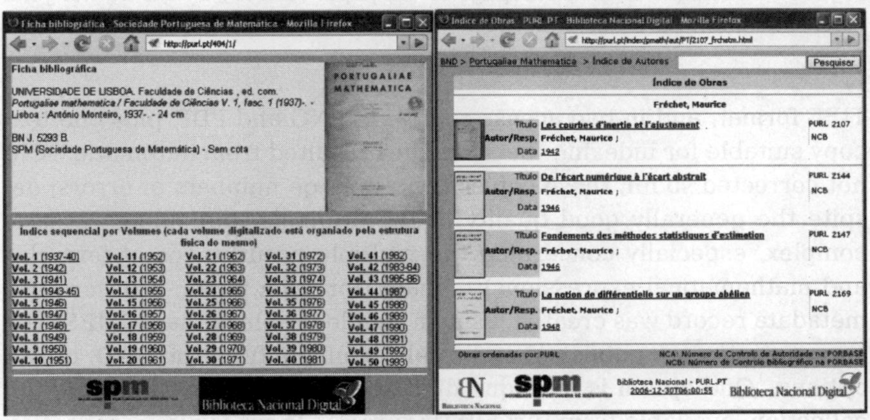

Figure 3. The homepage of *Portugaliæ Mathematica*, published as a unique digitized object, and an example of the index of papers by the author Maurice Fréchet.

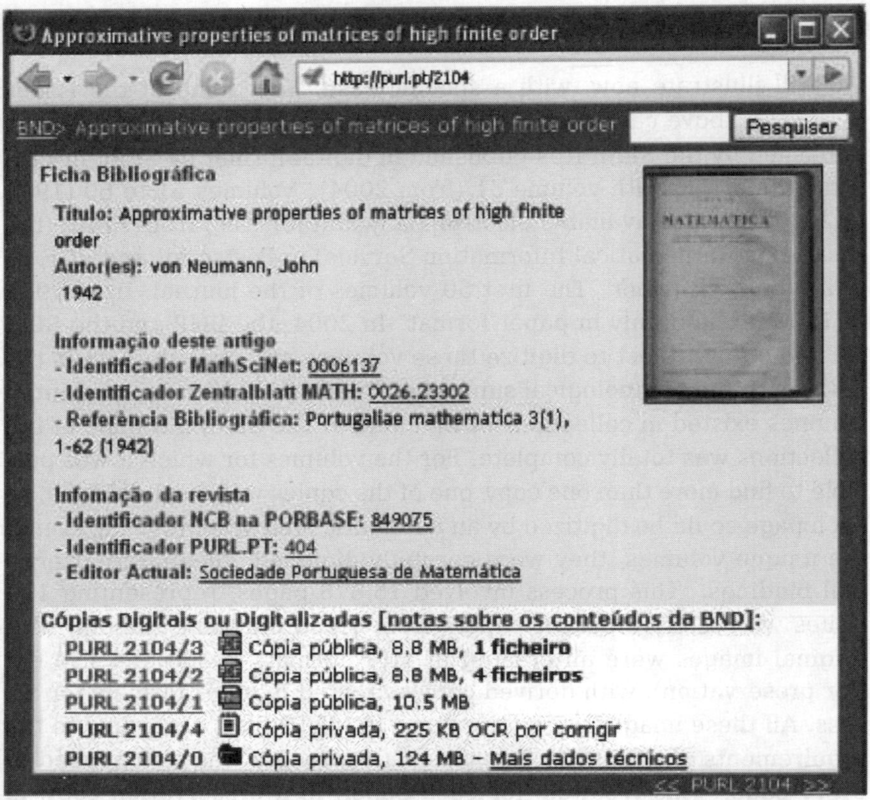

Figure 4. Example of the homepage of the paper [url:49], by the author John von Neumann, with links to MathSciNet and Zentralblat MATH.

TIFF format, and in two access copies, in PNG and PDF, plus one text copy suitable for indexing (because they resulted from automatic OCR, not corrected so far, these copies contain large numbers of errors; despite the generally good quality of the originals, the subject area is complex, especially concerning the multiple occurrences of formulas and mathematical expressions). Finally, for each paper, a descriptive metadata record was created, using a simple profile of the DCMES. The access copies are published and freely available from the BND, in two options. One option is a unique digitized object representing the full collection, available from [url:50] (Figure 3). There is also a collection of one digitized object for each paper, which makes it possible to reference each paper individually, as independent intellectual objects. This collection is available for browsing from [url:51], with several indexes, including indexes for each author as represented in Figure 3. Each paper in this option is also linked to its related description at the external services MathSciNet and Zentralblatt, as shown in the example of Figure 4 (each paper also has a link to the bibliographic record of the journal in PORBASE [url:45], the national bibliographic database). The descriptive metadata for each paper is coded in a profile of the DCMES. This metadata is available from OAI-PMH, from the server at [url:44], which is used by external services such as TEL and Mini-DML, which can retrieve the metadata coded as in Figure 5. The same metadata is also provided to Google Scholar, in the format represented in Figure 6. Finally, the same objects are available for direct searching in the BND by a Z39.50 service available at the DNS address <repox.porbase.org> (port “210”; database “bnd”; charset “ISO-8859-1”), which is used, for example, by the national scientific portal B-ON [url:72].

```
<OAI-PMH xsi:schemaLocation="http://www.openarchives.org/OAI/2.0/OAI-PMH.xsd">
 <responseDate>2006-12-30T15:33:02Z</responseDate>
 <request identifier="oai:oai.bn.pt:bnd/portmat2104" metadataPrefix="tel"
  verb="GetRecord">
  http://oai.bn.pt/servlet/OAIHandler
 </request>
 <GetRecord>
  <record>
   <header>
    <identifier>oai:oai.bn.pt:bnd/portmat2104</identifier>
    <datestamp>2006-04-17</datestamp>
   </header>
   <metadata>
    <dc>
     <dc:identifier xsi:type="dcterms:URI">
      http://purl.pt/2104
     </dc:identifier>
     <dc:title>
      Approximative properties of matrices of high finite order
     </dc:title>
     <dc:creator>von Neumann, John</dc:creator>
     <dc:description>Anal\'{\i}tico</dc:description>
     <dc:publisher>
      Sociedade Portuguesa de Matem\'{a}tica
     </dc:publisher>
     <dc:date>1942</dc:date>
     <dc:type>material textual, impresso</dc:type>
    </dc>
   </metadata>
  </record>
 </GetRecord>
</OAI-PMH>
```

Figure 5. The record of the paper in Figure 4 as it can be retrieved from the OAI-PMH server at the BNP.

```
<?xml version="1.0" encoding="UTF-8"?>
<articles xmlns:xlink="http://www.w3.org/1999/xlink/">
 <article>
  <publisher>
   <publisher-name>
   Sociedade Portuguesa de Matem\'{a}tica
   </publisher-name>
  </publisher>
  <article-meta>
   <title-group>
    <article-title>
    Approximative properties of matrices of high finite order
   </article-title>
   </title-group>
   <author-list>von Neumann, John</author-list>
   <contrib-group><contrib contrib-type="author"/>
   <name>
    <surname>von Neumann</surname>
    <given-names>John</given-names>
   </name>
  </contrib-group>
  <num_libraries>1</num_libraries>
  <self-uri xlink:href="http://purl.pt/2104"/>
 </article-meta>
</article>
```

Figure 6. The same record as in the previous figure, as it can be retrieved by Google Scholar.

A Digital Library Framework for the University of Aveiro

M. Fernandes, P. Almeida, J.A. Martins, and J.S. Pinto

With the increase of the amount of information that is produced by institutions such as universities, digital libraries and institutional repositories are performing a crucial role in the preservation and dissemination of digital content. In this chapter, we will overview the history of digital libraries, analyze existing open-source software for the implementation of institutional repositories, and discuss the University of Aveiro's digital library.

1 Introduction

The amount of information available to the public has increased greatly in recent years, with the Internet serving as a vehicle for both its production and dissemination.

To tackle the task of searching for large volumes of data, several search engines have been developed, such as Google, Yahoo!, and AltaVista. These are not, however, information repositories in the traditional sense. They instead provide repositories of links to information providers. These engines collect as much information from the webpages as they can find and store it in an unstructured and semantically meaningless fashion. Although this simplifies matters for the average user, who has only to type a few words in a search form, it also restricts the user from performing more complex queries in specific content.

Digital libraries, on the other hand, store large amounts of well described data in a structured and well-organized model and, although that is not always the case, they generally rely on the Internet. Their goal is to direct users to electronic collections, which may offer unique thematic value to researchers, historians, and general audiences.

It should be noted that we employ the expression "digital library" in its more generic definition, comprising digital archives, museums, and all similar systems. Digital archives, for instance, differ from digital libraries (in the strict definition) in the sources of information (primary/unedited instead of secondary), organization of information (categorically rather than individually), and preservation (a primary concern in archives). We will not make such a distinction.

Universities, as institutions that promote knowledge creation and dissemination, are being encouraged to build digital libraries/institutional repositories. The mission of these systems is to provide the necessary technological infrastructure to store, preserve, and disseminate scientific and cultural information.

The University of Aveiro is currently developing a digital library consisting of several systems. Each of these systems stores different information and is specialized to catalog and index specific information.

In this chapter, we will briefly analyze the evolution of digital libraries. Next, we will describe the architecture and the systems that compose the University of Aveiro's institutional repository, with particular emphasis on SInBAD (Integrated System for Digital Libraries and Archives) and eABC (Bibliographic Archive for Scientific Production). At the end of the chapter, we present some conclusions and future guidelines that were identified during the implementation process.

1.1 History of Digital Libraries

In this section we present a brief overview of the importance, and evolution over the last century, of concepts related to digital libraries. This information is the basis for understanding the rapid growth of digital libraries, as well as the increase of research projects and funding for building new and better systems.

The importance of digital libraries. Digital libraries are important in three major aspects:

1. *Simplicity in the consumption of, and search for, information.* By categorizing and describing data in a meaningful manner, digital libraries can provide simple list and/or search operations so users can find documents. Information consumption is also simplified, since users can have documents displayed on their monitors, and easily browse through large collections.

2. *Democratization of access to information.* Before digital libraries arose, much information was stored in libraries or, in worst cases, in rooms with difficult or restricted access, in specific geographic locations. Interested readers would have to travel to those locations and there inspect the intended documents. Digital libraries make it possible for a user to easily and quickly access information all over the world. Researchers can access papers from other universities and immigrants can read newspapers from their homelands.
3. *Preservation.* Digital libraries are crucial to digital preservation in two ways: a) they stimulate the digitization of historic material that would otherwise deteriorate without copies, and b) they employ preservation mechanisms, by maintaining digital objects in usable formats, performing backups, and undertaking replications.

The evolution of digital libraries. Surprisingly, some of the concepts behind digital libraries, such as preservation, have been present for more than a century. Microfilm technology, a compact storage medium for paper documents, is reported to have been first used in 1870 during the Franco-Prussian War [106]. Later in the 1930s, when World War II threatened to destroy the archive of the British Museum, University Microfilms started the preservation of printed works on microfilm.

In 1945, Vannevar Bush [48] proposed a system called "memex," where ultrahigh-resolution microfilm reels were coupled to multiple cameras by electromechanical controls. The prophetic endeavor also introduced a concept similar to hypertext.

The first remotely accessible databases came online in the late 1960s. These early databases mainly dealt with legal, scientific, and government information [253]. CD-ROM and local databases appeared in the mid-1980s, allowing images to be stored and retrieved.

In 1989, Tim Berners-Lee proposed a global and distributed hypertext information exchange network, which would become the HTML (HyperText Markup Language) based Internet [31].

In 1994, the Library of Congress announced a National Digital Library, and Libraries Initiative, a research effort involving several universities in the study of digital libraries [253].

In 1995, Kahn and Wilensky [149] defined an architecture of distributed digital objects services. According to the authors, a digital library belongs to such a category: it is a repository of digital documents, properly and uniquely identified, and information about those objects, called metadata. Later, in 1997, Arms et al. [12] presented an architecture for digital libraries and identified their four main components: digital objects, identifiers, repository, and user interfaces.

With the maturing of the involved technologies, current digital libraries face more challenges outside the technical scope, namely, copyright and legal issues [148].

Technology impact. Internet growth and its degree of adherence is the single most important factor in the evolution of technology in information systems. It has become the favored media of production and dissemination of information. Millions of users connect daily to a network of unknown dimension.

The need for document preservation, along with Internet growth and the evolution of desktop software and hardware, has ignited a quest for mass digitization of historic material—printed (books, letters, etc.), photographic (photos, posters), video (VHS and Beta), and audio (vinyl).

On the other hand, it has accomplished a dramatic shift in how society functions. For instance, many private institutional and commercial publications are no longer created in paper—only electronic versions are produced. Companies and individuals are starting to rely solely on digital invoices, reports, correspondence, etc.

While it seems clear that having all this digital material makes it easier to access and distribute information, it also points to the fact that efficient and easy to use information management software is crucial. Without such software, searching for a document in a hard drive with millions of files becomes little different from looking for a piece of paper in a stack of documents.

Technology evolution. The first digital libraries, as general information systems, were monolithic applications that used proprietary data and description rules. With the evolution of the Internet, researchers, librarians, and software architects found a need for the standardization of information and protocols to simplify communication among systems and ease the understanding of external data.

XML (Extensible Markup Language) has been the de facto standard for describing and transmitting data for some years. It provides a text-based language whose main purpose is to facilitate the sharing of data across applications, platforms, institutions, etc. Due to its flexible and customizable nature, XML has been the basis for numerous standards, such as SOAP, WSDL, XHTML, RSS, and technologies such as Web Services, OAI-PMH, and BPEL.

Web Services provide a standard and interoperable means of machine-to-machine interaction, using a well known interface, based on SOAP and WSDL. Web Services allow the transparent communication of machines with different programming languages, platforms, and operating systems. It also allows the aggregation and consumption of information in a simple way. Despite the standard interaction, there is no standard for the data structures being used. Even when two digital libraries store data with the exact same schema and metadata, each system does not have a priori knowledge on how to access information from the other: which remote methods to invoke, what data structures are provided, etc.

Open access and open archives initiatives have become popular in recent years. The underlying philosophy of these initiatives is the availability of digital content free of charge. It commonly embraces the concept of self-archiving, by which researchers make available their own work. Particularly important for interoperability between digital libraries is the OAI-PMH [140] (Open Archives Initiative Protocol for Metadata Harvesting). This HTTP-based protocol defines how a data provider exposes its metadata to harvesters (other digital libraries, federation sites, etc.) by using clearly defined XML structures, thus eliminating the problem of a priori knowledge.

2 Development of the University of Aveiro's Institutional Repository

In late 2004, the University of Aveiro, funded by the Aveiro Digital 2003–2006 program, started to remodel its Internet sites and applications and develop new ones. Most sites produced at the institution by that time provided standalone services, but the paradigm was dramatically changed. The new applications to be produced—the library's site, departmental pages, user management services, and many more—had to integrate, and interoperate with, each other, thus creating a network of cooperative and complementary systems.

Under these new circumstances, each system is solely responsible for its own data and must provide a predefined set of services and data when another system makes a request.

2.1 Existing Services

Regarding scientific and cultural publications, the two systems that provide most of the bibliographic information of the University are eABC and Aleph.

2.2 eABC

eABC [227] is a bibliographic archive for scientific production and is used by departments in the University of Aveiro to maintain an updated index of the work of researchers and teachers.

eABC not only stores the bibliographic references of the work developed in the university, but also serves as an authority database. In this database, all known authors are stored along with a history of their affiliations, and on each paper, author information is linked with the authority's records. The system is therefore capable of generating annual production reports for a department or institute, maintaining an updated publications list for authors, and showing who published with whom (who published with an author, which departments published together, etc.).

eABC was remodeled to use the new centralized user management service and to provide Web Services from which other systems can process information.

Aleph. The university's library is currently using Aleph's [170] application for the management of the bibliographic entries from most of the existing documents available (not only in the library itself, but also in the archive, the multimedia center, etc.).

This system is not a digital library, since only bibliographic information (the metadata) is managed; the documents in digital format are not stored. Metadata is stored according to the UNIMARC standard.

3 SInBAD

Prior to 2004, the university's digital documents were scattered in multiple locations: department sites, teacher's personal pages, e-learning systems, and more. To aggregate these digital objects of multiple sources and types, the university developed SInBAD [url:56], an integrated system of a digital library, archive, jazz repository, and museum. Due to the diversity of subsystems it aggregates, the SInBAD repository stores highly heterogeneous digital materials, such as images, videos, and sounds.

3.1 Information Metadata

To achieve a higher integration among all of these subsystems, a simple metadata schema is used for the basic object description: DCMI. Although this schema is insufficient to comprehensively describe an object from a specific source, it allows the storage and transmission of common properties such as title, author, date, etc. Each object's description is then further refined using metadata standards specific for the corresponding data type.

3.2 SInBAD Subsystems

SInBAD is composed of four subsystems: Library, Archive, Jazz, and Museum. Each of these subsystems is composed of its repository, a website, and a Web Service (Figure 1).

The subsystems are completely independent from each other and from the overlying portal. Users may access SInBAD contents either by using the portal or the corresponding website.

Digital Library. In the Digital Library subsystem, SInBAD primarily handles printed material: books, theses and dissertations, magazines, and scientific articles. Metadata descriptions for these documents are complemented by descriptors from the DCMI DC-Library Application Profile.

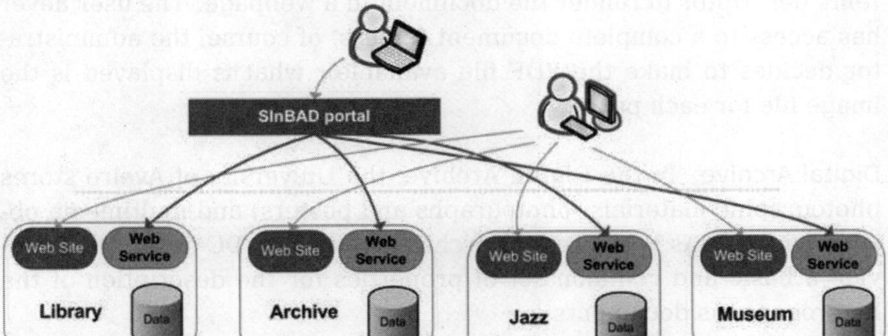

Figure 1. SInBAD subsystems.

To comply with restrictions in the delivery of copyrighted material, SInBAD uses the *tableOfContents* descriptor to subdivide documents into sections/chapters and to apply display and download limitations for each of those sections. By providing such a granular control of a document's access, the system can therefore provide administrators the means to make available only parts of the documents, limit the number of simultaneous accesses, and define temporal periods of availability.

Users may access one of these documents using one of two ways: by listing every existing title (sorted by title, author, or date) or by performing one of the available searches. For the printed material, SInBAD offers three search methods:

1. Simple, which performs an OR of the search query on the most common metadata descriptors (title, author, abstract, etc.);
2. Complex, which allows the user to specify what to search in each metadata descriptor;
3. Full-text, which allows the user to search for a specific phrase or expression in the documents content.

Despite the similarities in content, these types of documents are organized in distinct manners:

1. Books, theses, and dissertations are stored as objects sectioned in chapters.
2. Magazines are composed of a main entry (the magazine itself) and the child objects (the articles in the magazine).
3. Scientific articles are dealt with as a single object.

All of these distinct documents share a common display engine, which analyzes the description metadata and parses the *tableOfContents* descriptor to render the document in a webpage. The user never has access to a complete document (unless, of course, the administrator decides to make the PDF file available); what is displayed is the image file for each page.

Digital Archive. In the Digital Archive, the University of Aveiro stores photographic materials (photographs and posters) and multimedia objects (videos). As in the Digital Archive subsystem, DC descriptors provide a basic and common set of properties for the description of the heterogeneous documents.

For photographic material, SInBAD uses descriptors from the VRA-Core 3 [16] standard, which provides elements such as *measurements*, *material*, and *technique*. In the case of multimedia files, the standard employed is MPEG-7 [130], which allows a comprehensive description of a video and its segments.

As in the case of the Library, documents can be both listed and searched for and, in the latter case, both simple and complex (specific to the data type) forms are provided.

Digital Museum. The Museum subsystem holds museological data and metadata of the University collections. These currently include three collections: Iron, Ceramics, and Glass items.

The items in the collections are described using DC and a custom schema defined by the Portuguese Museum Institute [77], which allows the description of an item's historical track, employed technique, and composing parts, among several other descriptors.

Jazz multimedia repository. In the Jazz digital repository, there are stored albums (both from CDs and vinyl), musicians, magazines, and books about jazz. Unlike the other subsystems, the Jazz repository is a legacy database that was adapted to be integrated into SInBAD. It does not, therefore, use a description standard.

Currently, SInBAD allows users to pick and listen to any of the about 10,000 digitized tracks. There is also a radio available, automatically generated from randomly picked tracks.

3.3 Architecture

SInBAD's architecture is depicted in Figure 2. It is composed of data access modules, services and middleware components, and applications.

This modular architecture allows an easy decoupling of components. Services can be introduced in the network and new subsystems registered at the portal.

Only subsystems have a more rigid configuration, and are composed of a website, its middleware (which includes a Web Service), and a

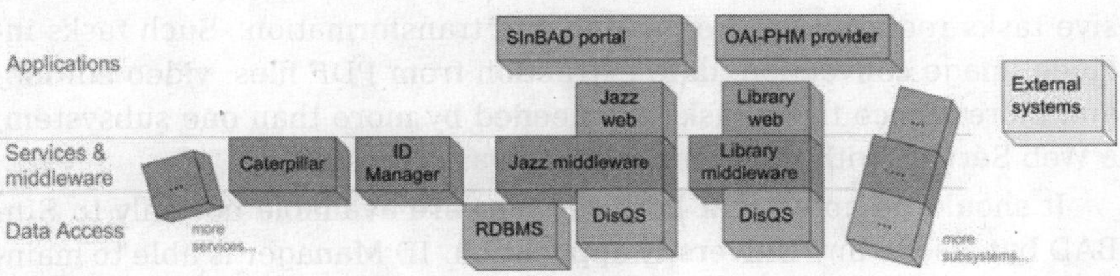

Figure 2. SInBAD's simplified architecture

DisQS module for handling data management. This configuration can, however, be changed without much effort.

Data access. The repositories are implemented using DisQS (Distributed Query System [120]), which allows contents to be distributed over several computers, as long as they have a DisQS Agent running. DisQS Manager, controlled by the middleware, virtualizes the access to the distributed repository as a single local disk, for instance by transparently aggregating searches. It is completely implemented with Web Services. DisQS also requires no relational database, since it is based on indexes generated from catalogs on the Agent's hard drive.

The Jazz subsystem is the only one requiring an additional component for data storage—a relational database (RDBMS)—since this is a legacy system.

3.4 Services

To meet the needs common to many applications and middleware, two services were created: ID Manager and Caterpillar. Both services are Web Services, thus capable of interoperating with virtually any platform and system.

In any collection of objects there is the requirement of unique identifiers that unambiguously identify documents, database entries, services, etc. In the RDBMS, this is usually achieved by defining a primary key column, which is automatically incremented to generate identifiers. In the decoupled subsystems scenario of SInBAD, however, repositories are distributed across several independent machines, but there is a need for identifiers that are globally unique. Also, by using DisQS, no relational database is used. With this in mind, we created a service that generates unique identifiers as they are requested by any subsystem: ID Manager.

Caterpillar service was created to address computationally intensive tasks required for file parsing and transformation. Such tasks include image conversion, data extraction from PDF files, video editing, and more. Since these tasks are needed by more than one subsystem, a Web Service with these capabilities was created.

It should be noted that both services are available not only to SInBAD but also to any University application. ID Manager is able to maintain separated identifiers for distinct scopes (applications), and Caterpillar is a general-purpose service.

OAI-PMH provider. Open Access is an important movement nowadays and the OAI-PMH is the de facto standard for interoperability among digital libraries and archives. The SInBAD portal has an OAI-PMH interface to expose its metadata to harvesters external to the university. SInBAD not only exposes its own metadata, but also metadata from eABC.

SInBAD portal. Sometimes, searching for a specific topic is more adequate than searching by a digital manifestation type. For instance, one may want to find information about some historical event (whether in books, posters, or videos) instead of narrowing it down to only one type of information.

To meet that requirement, SInBAD portal has a search form that simultaneously dispatches a query to the Web Services of the four subsystems. Results are aggregated, formatted in a coherent manner, and displayed to users.

3.5 Interoperability with Other Systems

SInBAD currently interoperates with eABC and establishes one way communication with Aleph (information flows from the latter to SInBAD).

SInBAD–eABC. As seen in section 2.2, eABC maintains an updated index of bibliographic references to scientific publications: articles, books, reports, and other material whose authors include University teachers and researchers. No digital copy of those publications, however, was available to users, since this was not the purpose of the system.

With the development of SInBAD, this situation has changed dramatically. Both eABC and SInBAD provide a Web Service by which the systems communicate.

Whenever an author or an institution anchor (on behalf of the authors) inserts a new publication on eABC, he may also upload the corresponding PDF file. This file, however, is delivered at SInBAD's Web Service, along with the eABC generated identifier, which processes it (with Caterpillar) and stores it in the Library's repository. If this operation is successful, eABC stores the URL for the document in its UNIMARC description file, in the format "http://sinbad.ua.pt/publicacoes/*identifier*".

When users search in eABC, each result whose digital manifestation is stored in SInBAD provides a link to it (in the above format). On the other hand, when users search for scientific publications in SInBAD, the system actually performs the search in eABC's Web Service, limiting the results to digitized entries. When accessing the digital document, SInBAD gets the UNIMARC description from eABC and displays it alongside the images (converted from the PDF).

Finally, SInBAD provides an OAI-PMH interface to eABC. It uses the eABC Web Service to obtain all digitized records, converts them to DC, and formats the output to OAI-PMH compliant XML.

SInBAD–Aleph. Since the bibliographic references for most of the existing material in the University are stored in this library's system, it would make little sense to ignore it and to undertake the time-consuming task of recataloging documents (or even copypasting the metadata).

With this in mind, every back office application we developed has an Aleph communication module. This module receives an identifier from the application (such as a system identifier or quota) and retrieves the corresponding UNIMARC metadata description from Aleph. This description is presented to the administrator (or editor) in the application's interface so that he can check the correctness of the information. When everything is validated and other tasks specific to the media are completed (such as specifying the table of contents and its permissions), data can be submitted. The back office will transparently translate the UNIMARC into DC or another description standard.

3.6 Conclusion and Future Directions

The development of a university's digital library (as an institutional repository) has a number of issues that are complex to deal with, not only at a technical, but also at a social, level. One must achieve interoperability with legacy systems and assure a minimum level of transition effort. It must also be in conformance with legal and social issues, such as copyright and privacy issues, and respect the institutions' hierarchy.

In this chapter, we presented SInBAD—the University of Aveiro's digital library—which, in conjunction with eABC, form its institutional repository. It successfully integrates with existing systems and gives access to digital content that had been inaccessible to the community: books, posters, photographs, videos, albums, etc.

Some important factors identified in the development of the University Institutional Repository are the modularity and the web services-based architecture that allow the future integration of new subsystems, services, and data providers. Also, its distributed topology offers greater scalability. As desired, SInBAD provides flexible and high granularity descriptions, highly configurable copyright protection, decentralized architecture, constantly updated indexes, and simplicity of installation.

As technology evolves, it is apparent that user input is becoming much more expensive and time consuming. There is therefore the need for automation of processes and machine-to-machine interaction. BPEL (Business Process Execution Language) [195] is an XML standard that allows the definition and execution of processes as a sequence of distributed tasks (Web Services) in an orchestrated manner. BPEL allows increasing automation, service reutilization, and productivity. Researchers at the University are studying the possibility of defining existing SInBAD operations as BPEL processes.

On the other hand, although SInBAD repositories are distributed, the control and coordination of operations are executed centrally. To achieve higher redundancy, fault tolerance, and scalability, the P2P (Peer-to-Peer) [272] paradigm is being considered. In P2P networks, each node/peer (a DisQS Agent, for instance) is both a client and server. The network is completely decentralized; it has no central point of control. Nodes therefore self-organize themselves in order to optimize network efficiency.

2 Technology Enhancements for Disseminating Mathematics

Coast-to-Coast (C2C) Seminar: Background, History, and Practice

J. Borwein, V. Jungic, D. Langstroth, M. Macklem, and S. Wilson

1 Introduction

The C2C Seminar (short for Coast-to-Coast[1]) is a seminar run jointly at universities throughout Canada, from Simon Fraser University in British Columbia, to the University of Calgary and the University of Saskatchewan in the West, to Dalhousie, Memorial, and other universities in the Atlantic Provinces. This seminar is simulcast to all sites via video-conferencing software, and each seminar provides opportunities for questions and comments from all of the remote locations.

The concept of the C2C seminar first originated with a large project called WestGrid. Starting in 2002, WestGrid was designed to be a massive parallel-computing infrastructure to be shared by eight Western Canadian universities, although this number has since expanded to 14 institutions as the project has gradually expanded eastward from B.C. and Alberta to also include universities in Saskatchewan and Manitoba. At the time, Simon Fraser University (SFU) also had an interactive lab and seminar environment called the CoLab (Collaborative Lab). This lab included a number of tiled touch-sensitive wall-mounted computer monitors, and was used for running courses and meetings, often remotely in cases where speakers were unable to attend the events personally.

[1] Along with all the technologies described here, our culture is also on the cusp of "text-speak." Thus many of us call the seminar the "Sea-to-Sea" seminar, the text-speak version of our name. This ironically produces an unintended alternative semantic [url:77]. (NT AL CHNGS R 4 TH BTTR!)

As WestGrid progressed, the goal was to have similar "grid-rooms" at each member university, to serve as local communication points for researchers who were working together on the WestGrid cluster from different institutions. In order to promote the resources that were available via WestGrid, a semi-regular event needed to be organized to show what could be done in terms of communication using this new infrastructure.

In late 2003, as WestGrid was built and began to populate its network with users from each member university, the CoLab research group moved to the Faculty of Computing Science at Dalhousie, to construct a new research environment called D-Drive (Dalhousie Distributed Research Institute and Virtual Environment), and with an additional goal of assisting ACEnet, a WestGrid-style shared network to connect universities throughout the Atlantic Provinces. During this same period, the CoLab environment at Simon Fraser was replaced by a much larger working environment called IRMACS (Interdisciplinary Research in the Mathematical and Computational Sciences). Once D-Drive and IRMACS were completed, the potential for a cross-Canada videoconference was obvious, and since 2005 the C2C seminar has enabled audiences from throughout Canada to attend lectures by distinguished speakers from across the country.

In this paper, we will discuss the structure of the C2C seminar, including the technical and organizational components, and the lessons we have learned during the start-up process. We will also provide details on how interested people can connect to this seminar from their own local university.

2 C2C Seminar: Structure and Content

2.1 Structure

The Coast-to-Coast Seminar is an hour-long presentation given on a topic from mathematics or computational science and made accessible to audiences at a number of remote sites through collaboration technology. Seminars are held every two weeks throughout the academic year alternating between the West Coast and the East Coast. Initially the Western and Eastern sites were IRMACS and D-Drive exclusively, but as the series grew, and included other universities, presentations in the series have also come from Edmonton and Calgary in the West, and from Acadia, St. Francis Xavier, and Math Resources Inc. (a Halifax-based educational mathematics software company) in the East. At the time of writing, presentations are also planned from the University of Lethbridge, Memorial University of Newfoundland, and University of New Brunswick, among others.

Audiences for a presentation are located at one or more discrete sites at universities across Canada. The collaboration technology enables two-way audio and video communication as well as a shared desktop. Thus a presenter is not only audible and visible to the audience, but can also respond to a raised hand, answer a question, or interact with an individual at a remote site through a shared application. The number of sites has increased to eight for an average presentation, with the promise of more participants in the future.

To set the stage for the presenter, we describe an outline of what a typical seminar entails. The actual presentation is expected to be of a high quality, yet accessible to a fairly general scientific audience. Accordingly, seminars are widely advertised and attract audiences beyond the realms of mathematics and computer science, depending on the presenter's topic. We emphasize that we may have three or thirty people at one or other of the sites, and that typically perhaps 60 to 80 people hear each of the talks. No one has to come just to ensure a respectable audience as is often the case in a departmental colloquium.

The main goal of the seminar is to give an opportunity to scientific communities from various Canadian universities to collaborate and share their interests. At the same time we aim to achieve several other equally important goals:

- To learn and understand the issues related to the organization and the running of a regular scientific event from several universities in different time zones.
- To set the standards for this type of event for the future.
- To test the available technology.
- To motivate the creation of new technological tools and to encourage the improvement of the existing tools.
- To give a chance to faculties to gain experience with presenting through a still relatively new medium.
- To educate the audience attending the seminar about the protocols and etiquette involved.
- To reduce the costs of inviting distinguished speakers.
- To justify the investment in the technology and in the people involved.
- To build a C2C community.

We aim for an environment that is no less familiar than a new seminar room. As Ron Fitzgerald crisply puts it, "No one has to explain chalk." That said, we follow a fixed protocol each time. Roughly 30 minutes before the seminar starts, designated individuals from each site confirm that all facilities are working at all sites. An introduction of all sites and of the speaker is made from the speaker's site. The speaker's presentation is approximately 45 minutes long and is followed by a question and answer (Q&A) session with all sites. The Q&A session starts with local questions and then rotates through the remote sites. As with a face-to-face seminar, the host determines when to stop—and a good host has a first question to start things off with.

2.2 Past Talks

The presentations to date in the Coast-to-Coast seminar have been a mix of mathematical and computational talks, with a wide variety of focuses within each field. As of the fall of 2006, the C2C seminar has featured the presentations listed in Appendix A.

During the summer of 2005, two test sessions were held. The first session consisted of several short presentations that ran from both IRMACS and D-Drive and were given by graduate and undergraduate students. The presenters were asked to use various methods and tools in delivering their talks: power point presentations, PDF slides, preprepared transparencies, writing on a white board, writing on paper and using a docucamera, using Maple applications, and so on. After summarizing the experiences from the first test sessions, a format for the future C2C presentation emerged. This format was tried during the second test session when Colin Percival from Simon Fraser University gave the presentation with the title "Hyperthreading Considered Vulnerable."

Following the success of the C2C Seminar Series over the 2005-2006 academic year, we hosted a more intensive distributed event, the Coast-to-Coast Miniconference on the Mathematics of Computation. This day-long event consisted of a series of six speakers, alternating between IRMACS, the University of Lethbridge, and D-Drive. The event was attended by audiences in each of these locations as well as in some of the other remote sites, according to interest and availability. Since then we have also experimented with shared open houses and other ways to experience "presence-at-a-distance."

2.3 Seminar Facilitation

Although we have been interested in producing a seminar experience for distributed audiences that is as close as possible to a face to face event, there are significant differences of which a local audience needs to be aware. In the first place, with increasing numbers of remote participants displayed on screen, it becomes more and more difficult for a remote audience member to attract the speaker's attention with a question. Thus the question period must be handled explicitly, with an active request for questions, usually at the end of the presentation. Even then, audience members need to be made aware that if they have been chosen to ask a question, they must wait until a wireless microphone is passed to them so that it will be clearly audible across the network.

The orientation of speaker to audience also requires a shift in the usual expectations. In the D-Drive lab, for example, the audience will see, facing them, the faces of all of the remote audiences displayed on the large monitors. A local speaker may choose to address both the live and remote audiences at the same time, shifting his attention from one to the other and standing in profile so that he can turn his head to make eye contact with the local audience or turning the other way to face his remote audience on the screen. Or, some speakers have chosen to sit facing the large monitors, speaking directly to the remote audience, but with their back to the local audience. Our experience has been that the former choice is usually the more successful, but we leave each speaker to find the orientation that they are most comfortable with.

In the hour leading up to the beginning of the presentation, conversations from one site to another across the technology link will usually be the business of the technicians, checking connections and levels. When everything is ready and the seminar begins, the microphone will be taken by either the Director or the Administrator at the site that is presenting the talk, in order to welcome audiences and to introduce the speaker.

For almost all of the speakers who have presented at the C2C Seminar, this was the first experience with giving a scientific presentation to an audience that was located in various locations of the country. Talking to the remote audience through an advanced but still very new technology is an additional challenge in communicating advanced scientific topics. One must appreciate the fact that the C2C presenters have been ready to take the risk by pioneering in the C2C experiment.

3 Technical Components and Issues

3.1 Technical Overview

The technology behind the Coast-to-Coast Seminar is a combination of open source software, standard PC hardware, and audio/video components. The structure is a client/server architecture, in which individual sites authenticate to a central coordinating server, with audio, video, and presentation data shared between all sites.

The seminar organizers chose to standardize on Argonne National Labs Access Grid (AG) software as the videoconferencing suite [3, 11, 61, 128, 172]. This selection was made for three primary reasons: 1) AG is quite flexible in site configuration, allowing full auditoriums with complex audio systems and multiple cameras to conference with individual PCs using webcams and headsets in the same collaborative session, 2) AG is platform independent, with clients available on Windows, Linux/Unix, and MacOS, and 3) AG is a highly scalable videoconferencing package, allowing up to 30-40 remote sites simultaneously (limited by available bandwidth and processing capabilities). Some examples of the use of Access Grid in remote collaboration can be found in [142, 180].

Access Grid sessions are coordinated through an Access Grid venue server, which builds upon a "rooms" and "lobbies" analogy to coordinate collaborative sessions. For the C2C Seminar, the WestGrid AG venue server is used as a meeting place for the virtual attendees, with the sites meeting in a virtual conference room on the server. Audio and video from all client sites are shared through the venue server, and the venue server also provides file sharing, presentation syncing, and limited chat capabilities. Information about the Access Grid layout in WestGrid can be found in [271].

Client sites provide information to the venue server about their audio and video capabilities, and this information is distributed to the other sites via the venue server. In the C2C Seminar, there are two primary client options: Access Grid clients and InSORS clients. The Access Grid client is provided by Argonne Labs as part of the Access Grid open source project. InSORS is a commercial extension of the Access Grid project that provides increased video quality and some additional collaboration tools, but is only available on Windows and MacOS. Each site uses whichever client the site finds most appropriate for the seminar series, and adjusts client settings for compatibility with the other sites. The C2C Seminar is dedicated to remaining platform independent due to the nature of the many different academic sites participating, and therefore the Seminar has standardized to toolsets that all platforms can use.

Desktop sharing from the presenting site (to view lecture notes, slides, and whiteboards) is provided via the open source Virtual Network Computing (VNC) software package. The C2C Seminar makes use of a VNC server product called VNC Reflector to create consistent connection-point and authentication information for all sites from week to week, with the presenting site delivering presentation information to the VNC Reflector via a VNC server. The VNC products are platform independent open source projects, in keeping with the Seminar's technical goal of maximum flexibility for connecting sites. To facilitate whiteboard viewing and presentation markup, the primary hosting sites (Dalhousie and SFU) provide SmartTech SMART Boards as drawing surfaces for their lecturers, which are then displayed over the VNC connection.

3.2 Sample Presentation Environment: D-Drive and IRMACS Layouts

The component technologies driving the Coast-to-Coast Seminar can best be shown by giving a description of the layout for one of the site locations for the seminar series, namely the D-Drive lab in the Faculty of Computing Science at Dalhousie University in Halifax, Canada. Although this research lab shows some of this technology in action, it would be a mistake to believe that one needs all of it in order to connect to the C2C seminar series—as mentioned earlier, one can connect simply with a web camera and the proper desktop software installed.

The D-Drive lab consists of five touch-sensitive computer displays, four cameras, several microphones, an echo cancellation unit, and a sound system, which together provide a suitable presentation and audience environment for a remote seminar. The displays, shown in Figure 1, consist of a centrally located 73" rear-projected monitor along with four 61" plasma screens, with two each on either side of the rear-projected monitor. These screens are separated into two sets of tiled displays as shown in Figures 2-3: the two left-most plasma screens are tiled as a single display, and are used for audio/video management and video feeds, while the three remaining screens are also tiled as a single display, and contain the presentation slides (on the central rear-projected display (shown in Figure 4) and various video feeds from other remote locations. Each set of tiled displays has a single point

Figure 1. The computer displays in the D-Drive research lab, one of the locations for the C2C Seminar series.

Figure 2. Left-pair of tiled plasma screens, used for audio and video management (left) and presenter's video-feed (right).

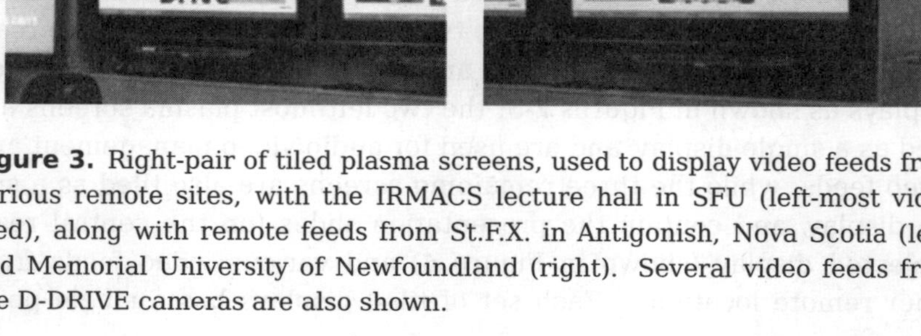

Figure 3. Right-pair of tiled plasma screens, used to display video feeds from various remote sites, with the IRMACS lecture hall in SFU (left-most video feed), along with remote feeds from St.F.X. in Antigonish, Nova Scotia (left) and Memorial University of Newfoundland (right). Several video feeds from the D-DRIVE cameras are also shown.

Figure 4. Central rear-projected display, with two mounted front-view cameras (top, cameras circled), with view from behind the left-camera (bottom-left image), and video-feed of left-camera (circled in bottom-right).

Figure 5. Camera views from back-left and back-right corners of the D-Drive research lab (left and right, respectively).

of focus, meaning that two users cannot simultaneously control different positions on different boards within the same tiled set of displays; therefore, the five displays were separated into two tiled sets in order to allow for local audio/video management and changes to video feeds (on the left tiled set) without taking control away from the presenter and his slides on the central display.

There are three locations for the audience within D-Drive to view a presentation: a conference table directly in front of the displays, seating area in the center of the lab, and tables along the back wall of the lab. The conference table is generally used for multi-site meetings, where there is no central presentation and where control of the discussion frequently changes between sites; for such events, the conference table has a centrally located microphone that picks up general discussion by anyone seated at the table. For most C2C seminars, the central seating area will be the primary location of the audience, with the conference table and the tables along the back wall serving as "overflow" seating when necessary. During the seminar, questions by audience members throughout the lab are asked via a wireless handheld mi-

Figure 6. Desk of Technical Supervisor for D-Drive, located to the right of the display environment, with contact with Technical Supervisors for all of the remote sites via messaging software on the desktop machine.

crophone that is passed to them when they indicate that they have a question.

The four cameras are spaced throughout the D-Drive lab, with two mounted on the top of the front displays (shown in Figure 4) and two more in the two back corners of the lab (shown in Figure 5). The cameras are placed in such a way that they give a sense of room context for remote sites, by providing multiple points of reference for the activities within D-Drive. The two front-mounted cameras provide the view of the local audience to the remote sites, with the left-camera being controllable by remote control in order to allow for panning and zooming when local audience members ask questions to the presenter, while the right-camera has a constant position that shows an overview of the audience seating.

The two rear cameras provide alternate views of the lab, and are primarily useful when the C2C presenter is speaking from D-Drive, or when someone is interacting with the SMART Board screens. In particular, the right-rear camera is also controllable by remote, and can provide a close-up video feed of the presenter to the remote sites, whereas the left-rear camera provides a wide overview of the screens and conference table. These cameras are not generally used when the presenter is located at another site, though they are quite useful for collaborative meetings in the facility.

In addition to the displays and the seating area for the audience, the D-Drive lab also has a desk located to the right of the display envi-

Figure 7. The presentation environment at IRMACS, for local (top) and remote (bottom) presentations during the C2C seminar.

ronment, where the local Technical Supervisor is located, as shown in Figure 6. This person controls the audio and video management on the left-most plasma screens, and is in contact with the Technical Supervisors at each of the other remote sites via instant messaging software on their respective local desktop computers. In case of technical problems, solutions are determined via discussion between sites, in part so that sites that are new to the seminar can get suggestions from more experienced sites.

The D-Drive presentation environment is just one of many within the C2C Seminar Series. To highlight some of the variety that exists within the various locations for the C2C series, Figures 7 and 8 show the display/presentation environment at IRMACS. This environment is

Figure 8. The locations of the desk of the Technical Supervisor at IRMACS (left-hand circle) and the rear videocamera, which is mounted on the back wall of the IRMACS lecture theater (right-hand circle).

contained within a lecture theater, with three touch-sensitive plasma displays and one or two projected displays. The presentation slides are shown on the larger projected display, as is the video feed for the presenter if located at a remote site. Figure 8 shows the location of the IRMACS Technical Supervisor within the lecture hall.

Although both D-Drive and IRMACS have extensive technology driving their respective display environments, there is no expectation that universities need to have a similar environment in order to take part in the C2C Seminar Series. In fact, one of the goals of the seminar series is to make it as easy as possible for universities to use their local space as presentation and display environments.

4 General Advice/Recommendations and How to Join

The Coast-to-Coast Seminar has provided stimulus for related uses of the technology, as we had hoped. These have included: distributed thesis defenses in which the examining committee is in more than one geographical location (for example, Thailand and Canada); shared planning meetings for academic projects; and joint seminars (e.g., Halifax/Ottawa/Brisbane for computer-assisted architecture). Many variations are possible—for example, during the 2006 Analysis Days [url:64] held at Dalhousie in January, Peter Borwein gave a plenary talk from

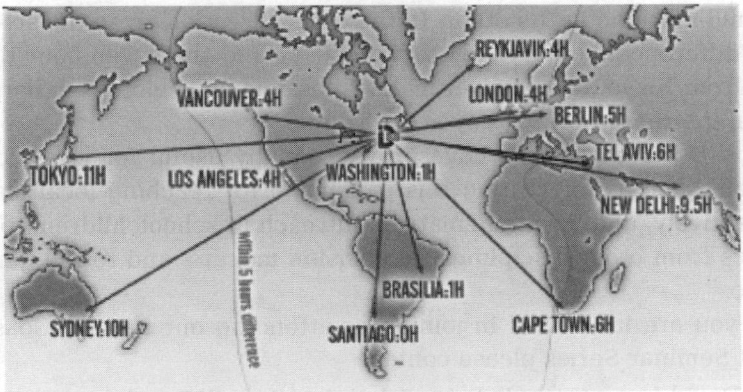

Figure 9. Time differences from D-Drive to various international cities (image courtesy of Andrew Shouldice).

his office at Simon Fraser University in British Columbia into the main auditorium at the Faculty of Computer Science at Dalhousie. Moreover, Jim Zhu participated in the entire two-day event from the Mathematics Room at Western Michigan University in Kalamazoo, while at the last minute Gabor Pataki gave his presentation from the University of North Carolina after he discovered that he could not get a visa in time.

The C2C Seminar currently focuses on research in the four Western provinces (British Columbia, Alberta, Saskatchewan, and Manitoba) and the four Eastern provinces (New Brunswick, Nova Scotia, Prince Edward Island, and Newfoundland and Labrador). The current absence of Ontario and Quebec is not intentional, and has arisen in part due to the nature of the presence of the Access Grid technology on the WestGrid and ACEnet networks. As similar networks are set up in Ontario and Quebec, we look forward to participation from universities from these two provinces[2].

Although the discussion of the C2C seminar has highlighted research collaboration within Canada, the presentation environments on both the east and west coasts are conveniently located for international collaboration as well. Figure 9 shows the time differences from D-Drive to various international locations, including lines demarcating locations within a five-hour time difference, which is easily sufficient to organize convenient joint meetings that will occur late-afternoon in one location

[2]The Coast-to-Coast Seminar is also not intentionally limited to Canadian universities. We welcome participation from American universities as well, provided that their participation is technically and logistically feasible.

and mid-morning in the other. In the other direction, even the six-hour time difference from Sydney to Vancouver and the seven-hour difference from Tokyo to Vancouver make it possible for remote collaboration within the same working day at both sites.

The D-Drive facilities have proved equally useful for local collaboration, such as for writing this paper and for teaching local classes interactively; and for mathematical outreach to schoolchildren, to academics from other disciplines, to decision makers, and to the general public.[3]

If you are interested in joining or attending our regular Coast-to-Coast Seminar Series please contact:

- David Langstroth, D-Drive Administrator, dll@cs.dal.ca
- Veselin Jungic, Associate Director, Research; at IRMACS, vjungic@sfu.ca

[3]All such uses are only as good as their weakest link, including such mundane technologies as the one controlling the security doors at the D-Drive facility, which have occasionally malfunctioned before seminars.

Digitally Enhanced Documents

K. Kanev, N. Mirenkov, and N. Kamiya

In this chapter, we propose a method and corresponding technological means for the digital enhancement of existing materials in mathematics and related subjects with additional content and functionality that could enhance comprehension by bringing in self-explanatory components, multiple views, and information sources on demand. Existing technologies for linking digital information to paper-based documents are analyzed and direct interaction methods and corresponding implementations are investigated. Applicability of the innovative Cluster Pattern Interface (CLUSPI®) for embedding digital content and references to digital texts, images, sound, video, and other multimedia content into traditional materials for math studies and research is evaluated.

1 Introduction

Mathematical documents tend to be quite information intensive and special measures have to be taken to ensure proper communication of their content. Along with document layout and typesetting, rigorous high standards related to document structure, content organization, notations, etc., must be followed [215]. In this chapter, we discuss innovative methods and technologies for digital enhancement of mathematical documents and the advantages that the rigorous organization and presentation of mathematical content could bring in this context. A recently published work of one of the authors [94] will be used throughout this article for illustrating our ideas and for providing practical examples. The full text of this work is available from the arXiv e-print

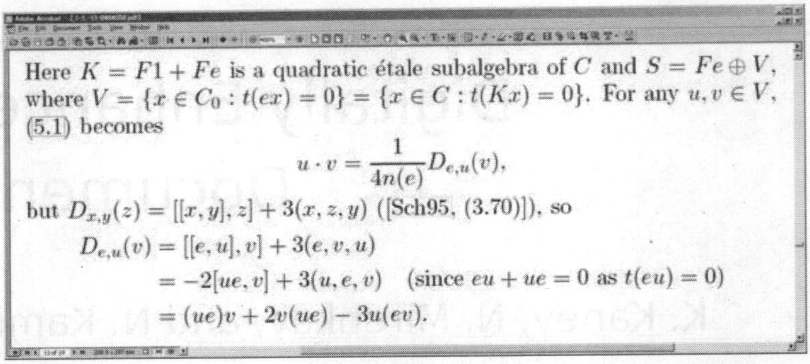

Here $K = F1 + Fe$ is a quadratic étale subalgebra of C and $S = Fe \oplus V$, where $V = \{x \in C_0 : t(ex) = 0\} = \{x \in C : t(Kx) = 0\}$. For any $u, v \in V$, (5.1) becomes

$$u \cdot v = \frac{1}{4n(e)} D_{e,u}(v),$$

but $D_{x,y}(z) = [[x, y], z] + 3(x, z, y)$ ([Sch95, (3.70)]), so

$$\begin{aligned} D_{e,u}(v) &= [[e, u], v] + 3(e, v, u) \\ &= -2[ue, v] + 3(u, e, v) \quad \text{(since } eu + ue = 0 \text{ as } t(eu) = 0) \\ &= (ue)v + 2v(ue) - 3u(ev). \end{aligned}$$

Figure 1. A sample PDF file of a mathematical document shown on a computer display.

service [url:5] in PostScript, PDF, and other file formats [94] and also as a full-sized journal paper [94]. Printed versions and screen versions of the work may significantly differ depending on the chosen file format and other factors. The PDF file, for example, when shown on a computer display appears as in Figure 1, while the printed version appears as in Figure 2. A close look at Figure 1 reveals two rectangles enclosing references to different parts of the text. The first rectangle, enclosing the string "5.1," is associated with formula 5.1 that appears earlier in the text. The second rectangle, enclosing the string "Sch95," is associated with the text of the corresponding bibliographical reference [229] in the end of the document.

Referenced parts of the text can be easily recalled by simply pointing and clicking on the corresponding rectangular area. This kind of linked referencing, however, seems to be only supported within the

Here $K = F1 + Fe$ is a quadratic étale subalgebra of C and $S = Fe \oplus V$, where $V = \{x \in C_0 : t(ex) = 0\} = \{x \in C : t(Kx) = 0\}$. For any $u, v \in V$, (5.1) becomes

$$u \cdot v = \frac{1}{4n(e)} D_{e,u}(v),$$

but $D_{x,y}(z) = [[x, y], z] + 3(x, z, y)$ ([Sch95, (3.70)]), so

$$\begin{aligned} D_{e,u}(v) &= [[e, u], v] + 3(e, v, u) \\ &= -2[ue, v] + 3(u, e, v) \quad \text{(since } eu + ue = 0 \text{ as } t(eu) = 0) \\ &= (ue)v + 2v(ue) - 3u(ev). \end{aligned}$$

Figure 2. A sample partial mathematical document printout from a PDF file.

scope of an individual document. In contrast to the display version, the printed version of the document in Figure 2 appears as plain text with no links at all. The rectangles marking the clickable links on the display seem to be automatically removed when the document is printed on paper. This kind of PDF file handling is understandable, given the fact that clickable links on paper are not supported.

In this chapter, we will discuss methods for extending linked referencing beyond the individual document limits by building document trees with reference branches that contain selective information from multiple documents. We will also introduce an innovative approach [152, 153] that bridges electronic documents and paper documents and enables readers to switch between different media in an easy and natural way. Paper documents with direct point-and-click functionality [154] will be considered as vehicles for maintaining clickable links on paper. We will also discuss the current system of signs and abstraction and how multimedia technologies can simplify comprehension of formal materials and communication between people.

2 Document Trees and Reference Branches for Mathematical Texts

Nowadays, almost every scientific journal offers online access to digital content either on a subscription or on a per article basis. Linking across document borders and ensuring transparent access to multiple documents content is not so much a technical challenge but rather a copyright issue, related to proper management of user access rights.

Typical references in science works usually direct the reader to an article, to a book chapter, or even to an entire book. Although a reader is probably going to need only a fraction of the cited work, because of the nonspecific reference, he has to obtain access to the full-sized original work. With this current system of referencing, each document can be considered as a root of a document tree that contains all the referenced documents in its branches.

References in mathematical works, however, tend to be much more specific and usually point to a single formula, lemma, theorem, etc. To illustrate this, let us consider the example shown in Figure 3. In contrast to Figure 1 where we had a nonspecific link pointing to the bibliographical data of the entire work in the end of the document, here we establish a link to the specifically referred formula 3.70 of the cited doc-

Here $K = F1 + Fe$ is a quadratic étale subalgebra of C and $S = Fe \oplus V$, where $V = \{x \in C_0 : t(ex) = 0\} = \{x \in C : t(Kx) = 0\}$. For any $u, v \in V$, (5.1) becomes

$$u \cdot v = \frac{1}{4n(e)} D_{e,u}(v),$$

but $D_{x,y}(z) = [[x, y], z] + 3(x, z, y)$ ([Sch95, (3.70)]), so

$$\begin{aligned} D_{e,u}(v) &= [[e, u], v] + 3(e, v, u) \\ &= -2[ue, v] + 3(u, e, v) \quad (\text{since } eu + ue = 0 \text{ as } t(eu) = 0) \\ &= (ue)v + 2v(ue) - 3u(ev). \end{aligned}$$

Figure 3. An example of a highly specific cross-document reference in a mathematical text.

ument. As can be seen in Figure 4, the referred text consists of a single line formula that represents a very small fraction of the more than 150 pages of content of the cited work [229]. In some cases, however, the same work may appear in different contexts and many separate citations referring to different parts of the work may be included. This is particularly true for voluminous books that play an important role in shaping a specific scientific field and thus tend to be heavily cited. Yet immediate acquisition of such material may be difficult, since books are rarely made available for instant downloading and it certainly takes time for a book to be physically delivered to the requester. Our point is that despite the increasing availability of mathematical documents online, immediate access to specific references might still be difficult.

We address this problem by suggesting that instead of document trees, reference trees containing only specifically required document parts could be used for mathematical documents. While building and distributing document trees of full-sized copyrighted documents might be problematic under the current copyright rules, building reference branches that only contain small, specifically required parts of cited documents might be permissible. Therefore, it might be beneficial for the mathematical society to establish a system for selective dissemination of, and access on demand to, limited portions of mathematical works that are directly referenced in other works.

$$(3.70) \qquad D_{x,z} = R_{[x,z]} - L_{[x,z]} - 3[L_x, R_z].$$

Figure 4. The specific formula from [229] directly referenced in the sample text of Figure 3.

3 Mathematical Documents in Digital Form

Along with the many mathematical journals and conference proceedings that are made available both in printed and in electronic form by their respective publishers, initiatives to expand online access are taking place in different countries. On the French MathDoc [60] portal [url:38], for example, access to bibliographical data and mathematical works is provided within the scope of the NUMDAM (digitization of ancient mathematics documents) and LiNuM (digitized mathematical books) initiatives. Similarly, the Goettingen portal [url:26] provides access to digital collections resulting from the retrodigitization initiative at the Goettingen State and University Library [210]. Global initiatives to support the long-term electronic preservation of mathematical publications worldwide such as WDML [141] and EMANI [269] are also being pursued.

According to our knowledge, current initiatives are focusing on the digitization, storage, and retrieval of full-sized mathematical documents. We believe, however, that some of the above-mentioned existing services could be used as a basis for building electronic depositories with selective partial mathematical document content, as we will discuss in the following sections.

Access to mathematical content in electronic depositories can take different forms. For example, a reader can use a computer connected to the Internet to retrieve referenced content and display it on a computer screen. Alternatively, the reader may wish to work with a printed copy of the retrieved electronic document. Since appearance and functionality of the screen and the printed versions of the electronic document are quite different, readers may perceive them as loosely connected or even as independent. Both document versions, however, are based on the same original electronic document from the depository. In this context, the choice of digital formats for document representation in the digital depository, for screen presentation, and for printing play an important role.

Ideally, screen presentation should be handled through the standard World Wide Web browser functionality. However, due to the complex mathematical notations that are an essential part of any mathematical document, special fonts and layouts that go beyond standard browser capabilities are often needed. Despite that, quite useful solutions exist, for example, the TtH translator software package [url:30], which converts plain TeX and LaTeX documents to HTML. Another approach is to use embedded graphics in HTML. This method is adopted by the

LaTeX2HTML [125] converter software [url:61], which uses small images to represent equations. More recently, the MathML standard [url:186] for representing mathematics in electronic documents has been developed. Although some browsers provide direct native support for MathML, others still need external plugins [54].

With all of the above-mentioned approaches, document appearance in a browser is much different from its printed version. Different document appearances on different media may actually help improve document readability and could provide better support for specific document functionality associated with different media. For example, while interactive use of mathematical documents on computer screens is quite common, for paper documents it is still to come. The work [67] addresses issues of interactive mathematical documents on the web in more details. It is based on the OpenMath standard [url:153] for communicating mathematics across the web and focuses mainly on expression of mathematical content. This is in contrast to the earlier-mentioned MathML standard, which mainly focuses on presentation.

Other methods are designed to ensure consistency of document appearance on different media. Typical examples are document formats such as PDF and PostScript, which can be used both for browser visualization and for printing through specialized external plugins [134].

4 Mixed Use of Digital and Printed Mathematical Documents

Functionality of a depository of full-sized mathematical documents does not need to be much different from all existing online preprint and article services. When we are considering an electronic depository with selective mathematical content, however, access functionality and corresponding user interface may need to take a different form. First of all, the selective mathematical content is only valuable within the scope of a specific reference of some initial mathematical work source. If the initial mathematical text is available in a digital form, methods discussed in the previous section can be applied. In this section, however, we will consider cases when the initial mathematical document is available in printed form.

Obviously, a reader can use a computer connected to the Internet for retrieving the referenced content and displaying it on the screen, but usually it is perceived as inconvenient to put down the original

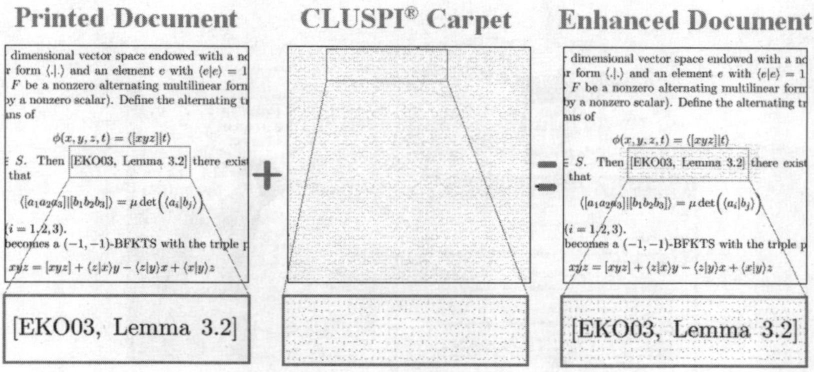

Figure 5. The Cluster Pattern Interface (CLUSPI®) method illustrated.

printed document and then to turn to the keyboard and the screen for retrieving referenced materials. Even after retrieving the material, perusing it on the screen may be difficult if one needs to switch between several references. A printed copy might serve much better the user's needs by allowing flipping of pages and even taking handwritten notes.

We are addressing the above-outlined issues by proposing a system that seamlessly integrates printed documents and electronic repositories by providing instant access to digital documents through paper-based point-and-click functionality. Our method is based on the Cluster Pattern Interface (CLUSPI®) technology [152, 153] coinvented and patented by one of the authors. With this technology, nearly invisible digitally encoded layers of clusters can be created and overlapped on document surfaces [154]. Resulting cluster patterns are specifically designed to be unobtrusive and still to provide a robust and reliable decoding and recognition. The digitally encoded layers become an integral part of the printed document and can be processed and printed along with the standard document content while staying practically indistinguishable to the human eye. An illustration of the CLUSPI® method is shown in Figure 5.

When a special CLUSPI® device is pointed to the digitally enhanced document surface, the cluster pattern layer is separated from the document content and is digitally processed. As a result, the pointing device position and orientation in respect to the document surface can be derived. Based on this data, direct access to the referred digital content in electronic form can be effectuated, as illustrated in Figure 6.

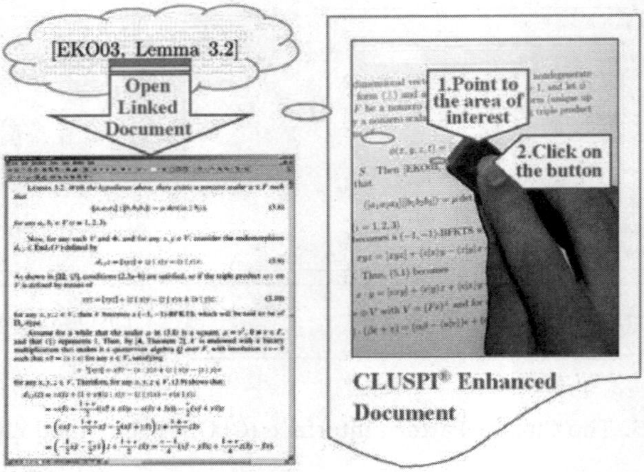

Figure 6. Using CLUSPI® enhanced printed materials for digital document access.

An important difference between traditional interfaces for access to digital documents and the CLUSPI® direct point-and-click interface is that no other input devices such as a mouse or a keyboard are required. Please note that when a mouse is used, cursor feedback on a computer display is mandatory. With the CLUSPI® direct point-and-click functionality, however, no cursor and therefore no display feedback is needed.

One can easily envisage now a CLUSPI®-based system that uses paper as a sole medium and has no keyboard, no mouse, and no display. With such a system, once the user clicks on a reference in the work being read, the corresponding electronic document can be instantly accessed and its referred parts can be sent to a printer. Having a printed copy at hand, the reader can flip the pages and peruse the document content as many times as needed. Additional advantages of the paper medium are possibilities for writing down notes, filing the original publication together with all references in printed form, and others. Referenced digital documents along with the standard content can also contain CLUSPI® layers, and therefore once printed can again be used for direct paper-based interactions. This way multiple levels of references can easily be supported.

Further extensions of our method could be achieved by introducing access to self-explanatory components [281] and additional dynamic content such as video and (possibly) sound. Obviously, dynamic images cannot be directly printed on paper so key frames and other visualization techniques are usually applied to convey some general idea of the video content. The real video playback, however, remains still essential if complex phenomena need to be studied and understood. With the CLUSPI® ability to directly link any digital content to a printed document, direct point-and-click on a key frame, for example, may initiate video playback from that specific key frame. A CLUSPI®-based implementation could follow the "video paper" concept [137] for a paper-based video browsing interface. The video paper conspicuous barcodes, however, would be replaced with the unobtrusive CLUSPI® carpet encoding. This way, no dedicated area for barcodes would need to be reserved on the printed document surface.

5 Experimental Implementation

Our experimental implementation is based on [94] and the references that appear in it. First, an electronic version of the original document in PDF format has been obtained and all external references have been identified. Specifically referred parts of the external documents have been extracted from existing PDF files or scanned from paper. Resulting digital content have been converted to PDF and appended to the original document source. This way, a single PDF file with the entire original document content and the specifically referred parts of the other documents has been produced. The resulting PDF document represents a simple, single-level document that can be browsed by clicking on the included specific reference links (Figure 3).

Our next step was to enhance the resulting PDF file with direct point-and-click functionality. For this purpose, CLUSPI® digital data has to be added to the PDF file in such a way that, when printed, clickable areas would appear around the references in the document. The reference map shown in Figure 7 has been designed to accommodate the external references and additional clickable areas. On the reference map, each sensitive area is represented as a rectangle, surrounding the corresponding text. Every such rectangle defines a portion of the CLUSPI® carpet that has to be cut and then overlaid on the document page accordingly (Figure 5). Please note that, based on the position and orientation of the CLUSPI® input device, multiple links can be associated with a single rectangular area. In this manner, for example, the names of the three authors in the top area in Figure 6 are linked to their respective World Wide Web sites.

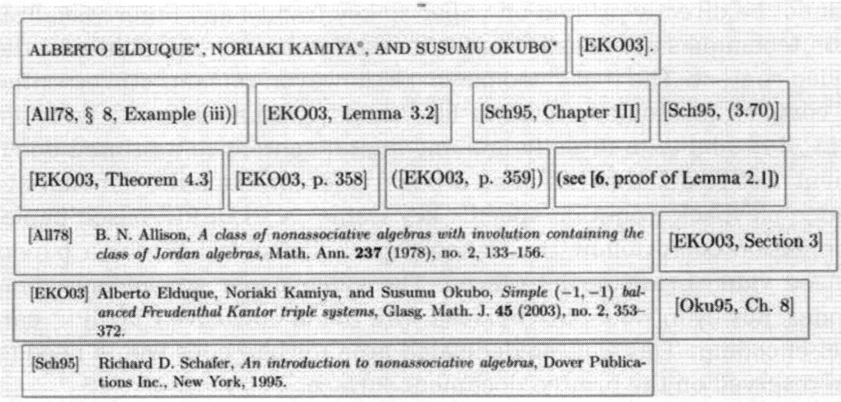

Figure 7. A reference map with CLUSPI® encoded areas.

The unobtrusive character of the CLUSPI® code helps preserve the original appearance of the mathematical document both on a computer display and on paper (Figure 6). According to our experience, however, some users would prefer to be able to clearly distinguish and see the clickable areas of the document. To accommodate this we have added an optional highlighting facility that can be applied to specific document areas. A partial view of a real document page with some highlighted CLUSPI® encoded areas is shown in Figure 8. CLUSPI® enhanced electronic and printed documents developed for our experiments represent building blocks of a broader structure covering different aspects of mathematical research and education as shown in Figure 9. We are currently working on digital enhancement and integration in this structure of additional documents and reference materials. The chart in Figure 9 itself can be printed on a standard A4-sized sheet of paper and then used as an interactive interface medium for point-and-click access to such materials. Direct online access to relevant web documents can also be provided through a CLUSPI®-enabled computer with an Internet connection. In our experimental implementation, for example, the first from the top rectangle in Figure 9 is linked to the previously described single-level document reference tree in PDF format. Similarly, the third from the top rectangle is linked to a mathematics subject classification page on the web that contains references to additional subject specific information.

CLUSPI® enhanced documents, linked materials, and interface components could be used for very practical purposes. For example, a

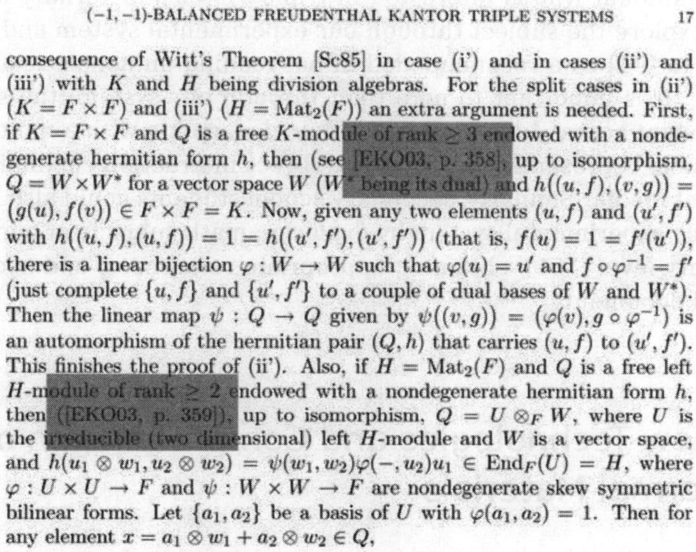
(−1, −1)-BALANCED FREUDENTHAL KANTOR TRIPLE SYSTEMS 17

consequence of Witt's Theorem [Sc85] in case (i') and in cases (ii') and (iii') with K and H being division algebras. For the split cases in (ii') ($K = F \times F$) and (iii') ($H = \mathrm{Mat}_2(F)$) an extra argument is needed. First, if $K = F \times F$ and Q is a free K-module of rank ≥ 3 endowed with a nondegenerate hermitian form h, then (see [EKO03, p. 358]), up to isomorphism, $Q = W \times W^*$ for a vector space W (W^* being its dual) and $h\big((u,f),(v,g)\big) = \big(g(u), f(v)\big) \in F \times F = K$. Now, given any two elements (u,f) and (u',f') with $h\big((u,f),(u,f)\big) = 1 = h\big((u',f'),(u',f')\big)$ (that is, $f(u) = 1 = f'(u')$), there is a linear bijection $\varphi : W \to W$ such that $\varphi(u) = u'$ and $f \circ \varphi^{-1} = f'$ (just complete $\{u, f\}$ and $\{u', f'\}$ to a couple of dual bases of W and W^*). Then the linear map $\psi : Q \to Q$ given by $\psi\big((v,g)\big) = \big(\varphi(v), g \circ \varphi^{-1}\big)$ is an automorphism of the hermitian pair (Q, h) that carries (u,f) to (u',f'). This finishes the proof of (ii'). Also, if $H = \mathrm{Mat}_2(F)$ and Q is a free left H-module of rank ≥ 2 endowed with a nondegenerate hermitian form h, then ([EKO03, p. 359]), up to isomorphism, $Q = U \otimes_F W$, where U is the irreducible (two dimensional) left H-module and W is a vector space, and $h(u_1 \otimes w_1, u_2 \otimes w_2) = \psi(w_1, w_2)\varphi(-, u_2)u_1 \in \mathrm{End}_F(U) = H$, where $\varphi : U \times U \to F$ and $\psi : W \times W \to F$ are nondegenerate skew symmetric bilinear forms. Let $\{a_1, a_2\}$ be a basis of U with $\varphi(a_1, a_2) = 1$. Then for any element $x = a_1 \otimes w_1 + a_2 \otimes w_2 \in Q$,

Figure 8. Snapshot of a document page with highlighted CLUSPI® encoded areas.

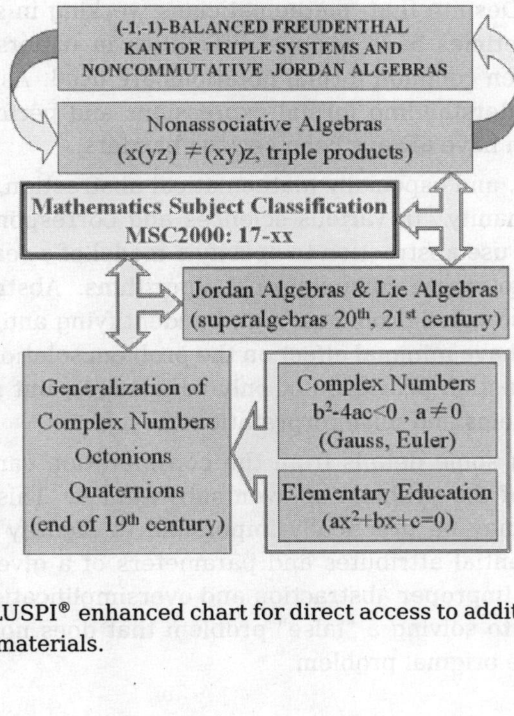

Figure 9. A CLUSPI® enhanced chart for direct access to additional documents and reference materials.

master student who is interested in triple systems or ternary algebra could explore the subject through our experimental system and obtain access to [94] as a recommended initial reading on nonassociative algebras. It is important to note that, through our system, the student gets instant access to generic electronic documents with a multiplicity of possible representations retaining direct interaction capabilities on different media. Similarly, secondary school students could also benefit from our experimental system by exploring math subjects at a more elementary level [151], obtaining appropriate references, and accessing related documents.

6 New Technologies and Comprehension of Formal Materials

Mathematicians are a special category of people who can introduce formal notations, define abstract constructs, and perform manipulations with such constructs. They are capable of using very abstract languages and they continuously improve this capability in their everyday work. Despite that, mathematicians working in slightly different fields sometimes have serious difficulties in understanding each other, even when common formal notations are used. As for nonmathematicians, understanding formal expressions and performing operations with them have always been serious barriers.

Abstraction, and especially mathematical abstraction, is a great invention of humanity. In various sciences and corresponding applications, we often use abstraction to specify a model of a real-world problem and then to create computational algorithms. Abstraction allows us to simplify complex problems, e.g., by identifying and putting away all details that have minimal effect on the problem solution. Sometimes however, abstraction provides not only advantages, but may also be a source of problems and misinterpretations:

- Excluding some details from the consideration can obscure the essence of the problem or even substitute it. This happens because it may be practically impossible to identify correctly the most essential attributes and parameters of a given problem in advance. Improper abstraction and oversimplification may therefore lead to solving a "false" problem that does not consistently reflect the original problem.

- Absence of some data can bar us from finding efficient computational schemes or even the solution itself.
- Abstract things are hard to understand and abstractions can actually make representation of methods and data even more sophisticated, thus increasing communication barriers between people from different fields of activity. When one uses abstract sets of symbols, definitions, and functions, for example, others might perceive such abstractions as forming an encryption system that requires one to go through a decoding process before getting to the essence of the models or the methods being described.

Many existing abstractions and, specifically, mathematical abstractions, were developed quite a while ago. Similarly, the majority of current computer science abstractions are based on mathematical functions, text-type symbols and features of computers from 30-40 years ago. Since a fundamental point of any abstraction is its level, perhaps we should consider most of the existing abstractions, especially in computer science, as being of too low a level. We believe that new, higher-level abstractions are becoming essential in order to accommodate the rapid growth of computing power and new advanced computer features.

Since long ago, people have been carving and drawing pictures in order to represent objects and capture actions. Pictures were easy to understand, and for ancient people they were the natural techniques for representation of data and knowledge. Even with the poor technologies of the past, such an approach provided a way to decrease the abstraction level so that reasonable data recording and knowledge preservation could be achieved. However, perhaps because recording in pictures was still too time consuming, people drifted away from this self-explanatory style. Step by step, picture languages were replaced by languages where signs represented not objects and actions but syllables, phonemes, and, in some cases, words. As a result, most of the current writing systems are reflections of spoken languages rather than direct representations of data and knowledge. Probably there were a lot of reasons for such an evolution, but one of the main reasons might be the limitations of past technologies.

With the more advanced current technologies, however, we may be able to revert this evolutionary process back to pictures and images and even to augment it. Indeed, results of the sign system evolution might not look so attractive if evaluated in the light of more recent technological achievements. For example, existing gaps between sign expression syntax and corresponding semantics may be considered as a major drawback. In fact, we represent our data and knowledge by encoding them with systems of signs, and therefore we need to perform corresponding operations of decoding in order to understand that data and knowledge. Unfortunately, developing such encoding-decoding skills for different types of signs, including mathematical ones, is a very time-consuming process. Even after many years of education at schools and universities, people still spend a lot of energy and time struggling with these skills and techniques.

Motivated by the preceding discussion, we will be trying to introduce a new system of "signs" for bridging the gaps between syntax and semantics. Our approach is based on an idea of self-explanatory components [188] and is used as a basis for algorithm specifications, human-computer interfaces, and new types of educational materials. The components are created in a "cyberFilm" format, which is a new type of abstraction that combines both physical and mathematical concepts. Physical concepts are used to represent relations based on space, time, color, and sound. Many possible relations could be represented by such natural means and the role of conventional mathematical functions can be drastically decreased. A cyberFilm is a set of multimedia frames where each frame represents a view or an aspect of an object or a process. Every cyberFilm represents multiple views of an object with many aspects. Such a multiple view approach becomes an essential vehicle for introducing the concept of self-explanatory objects. A self-explanatory cyberFilm, for example, might have associated frames organized and presented in such a way that the semantic richness of data and knowledge is clearly expressed.

Detailed examples of cyberFilms related to sequential and parallel matrix multiplications, solving algebraic and partial differential equations, inter-component communications, as well as cellular automation-like algorithms and algorithms on trees and pyramids, can be found in [90, 91, 135, 222, 224, 263, 280, 281], respectively.

Our method arranges the frame views into six specialized groups. The first three groups visualize: 1) computational steps and data structures, 2) variables attached to the structures and formulas used in space-time points of algorithm activity, and 3) input/output operations. Each of the three groups can be observed and manipulated rather independently. The fourth group consists of frames of an integrated view where important features from previous groups are presented together. Groups 5 and 6 are the auxiliary views related to the film title, authorship, registration date, and links to other algorithms, additional explanations, statistics of the usage, etc. And last but not least there is

Figure 10. The cyberFilm format structure.

one additional group, Group 0, where meaning is presented as a whole through "what" it is instead of "how" it is. The cyberFilm format structure is depicted in more detail in Figure 10. CyberFilms as pieces of "active" knowledge [187] are organized in a film database. The frames are watchable and editable in a nonlinear order according to user demand.

The cyberFilm notation improves the mental simulation of application algorithms and gives greater assurance that the design of a method is correct. CyberFilms can provide more information than text, graphics, and images. The information related to the position, timing, distance, and temporal and spatial relationships is included in the cyberFilm data implicitly. Related change of objects between different frames provides a lot of information about the behavior of these objects. CyberFilms show how images, animation, and text are coming together to represent data and knowledge. The self-explanatory concept is also an abstraction; nevertheless, it is intuitively much more understandable and allows employing a multiplicity of "fuzzy" views to represent an accurate meaning.

The introduction of multiple view objects is a fundamental basis for developing next generation educational materials. Special tools should be developed to support the idea of active and easy to understand knowledge. The CLUSPI®-based approach presented in the previous sections is a promising technique allowing the attachment of additional aspects to existing pieces of knowledge through multimedia channels and, in doing so, increasing the comprehension of the knowledge.

7 Conclusions and Future Work

Learning mathematics is a step-by-step process in which advancement to the next step is only possible after mastering the preceding steps. Accordingly, education process in mathematics is structured in a way that encourages backward referencing and therefore repeated revisions of specific, previously introduced educational content. We believe that our CLUSPI®-based method would significantly increase document interactivity, facilitate the back-referencing process, and enable readers to focus on the content and its meaning rather than on the ways to get access to it.

Our approach has been tested so far with research publications that could be used for educational support at universities as well as for self-study and research. We are currently working on enhancing educational materials for elementary and secondary school math classes. The concept of a new generation of CLUSPI® enhanced educational materials based on multiple view objects that could be seamlessly integrated not only in digital but also in printed documents is under continuous development and new applications in elementary, secondary, and high school education and training are emerging.

Speech and Tactile Assistive Technologies

C. Bernareggi, V. Brigatti,
D. Campanozzi, and A. Messini

At present in Europe, visually impaired students seldom attend scientific courses because of the many difficulties in working with mathematical expressions and graphics by using tactile and speech tools. In the last decade, many screen reading technologies have been developed, so nowadays blind people can use many software applications for graphical user interface operating systems. Nonetheless, currently there is not a mathematics reading and writing system fully usable by both blind students and sighted assistants, which exploits recent information technologies and integrates with mainstream applications. Furthermore, there exist many tools to enable blind students to read tactile images, but few tools that enable blind students to create tactile graphics on their own.

Two tools that improve access to science through touch and auditive perception are introduced: the LAMBDA system and a Mathematica-based package to create tactile images.

The IST European LAMBDA project developed a mathematical editor for blind users. It exploits recent screen reading tools and braille displays to convey mathematical expressions. Many innovative working strategies, which aim at compensating for the lack of sight, are implemented in the LAMBDA editor. The overall system has been evaluated for the last year in many European schools and the results are extremely promising, especially with respect to usability. Moreover, this editor can be easily integrated with mainstream applications through the MathML format and a Python-based scripting language.

A study was conducted in order to produce a set of tools that simplify the preparation of tactile images. These tools are based on the

Mathematica calculus software. They were designed in order to avoid any visual interaction, so blind students themselves can create some classes of technical drawings (e.g., real function diagrams or figures for Euclidean geometry).

1 Speech and Tactile Access to Scientific Documentation

1.1 Scientific Documentation

Text, mathematical expressions, and graphics are the chief means used to describe and communicate concepts and procedures in scientific branches of learning. As for input, expressiveness, structure, and understanding, these documentation forms present peculiarities and differences.

Text is the printed representation of natural language. It can be written on paper by pencil and it is easy to input it in a digital form using a computer keyboard. Actually, a lot of editors and word processors exist and they have proven to be advantageous in many writing activities, including taking notes and preparing documents. Tools and techniques to work on text are widespread (e.g., text marking, cut and paste, and search and replace). Text is generally organized according to a rather simple structure (e.g., words, sentences, paragraphs, sections, and chapters), consequently, the exploration of its visual rendering can proceed either from left to right or by browsing the tree structure of the parts. The mental ability necessary to understand textual information is usually not high, as it is often enough to get the gist of what is written to fully understand the meaning of a whole sentence. Therefore, not all the written items have to be carefully read and remembered to properly understand a concept described through text.

Mathematical expressions are meant to describe unambiguously concepts and procedures. There exist many tools to input mathematical expressions in a digital form, but they are mainly used to transcribe something written on paper for printing or displaying on a screen. Writing on paper is still widely regarded as a more straightforward technique to take notes, to simplify algebraic expressions, and to understand symbolic manipulations. This is due above all to the difficulty in designing effective, efficient, and satisfactory input and edit techniques for mathematical notations exploiting two-dimensional representations (e.g., fractions, subscripts, superscripts, exponents, and matrices). Mathematical expressions are potentially extremely structurally complex. For example, a sum of nested fractions is easy as for operations to perform in simplifying it, but it is complex as far as the structure is concerned. In order to improve readability, the visual rendering of mathematical expressions exploits a two-dimensional layout. It induces horizontal and vertical exploration, as well as the necessity to access directly specific building blocks (e.g., the numerator and the denominator) precisely located on a two-dimensional plane. Graphical descriptions are related to a large variety of representations, such as diagrams, flowcharts, drawings, and so on.

The images are extremely expressive. An image can convey synchronously much meaningful information. Actually, visual exploration of images is not a sequential process like text reading. Many meaningful informative units are instantly isolated by sight and mental processes; afterwards mutual relations are constructed. For example, observing the diagram of a sinusoidal function, the Cartesian axes are immediately isolated from the curve, and mutual relations between the curve and the axes (e.g., intersections and position of some points in a certain quadrant) are rapidly understood.

The possibility to come effectively, efficiently, and satisfactorily to a mental representation of text, mathematical expressions, and drawings is a basic requisite to understanding scientific documentation. Therefore, attention is focused on these three components of scientific documentation.

1.2 Tactile and Speech Assistive Technologies

Tactile and speech user interfaces are generally designed for visually impaired people and they are commonly implemented in assistive technologies, namely, those tools that allow one to overcome difficulties due to a disability in achieving tasks (e.g., reading and writing documents, and browsing web pages).

Screen readers are programs that make the information displayed in a visual interface available in a different modality: either Braille through a Braille display or speech through a speech synthesizer. These programs are general purpose because they attempt to improve access to any information presented by programs within a particular operating system. This enables visually impaired people to use the same software products as their sighted colleagues. Nonetheless, interaction paradigms used in graphical user interfaces are sometimes unsuitable for, or even totally unusable by, visually impaired users. Therefore, specialized programs are produced to fulfill needs and operative strategies not catered to by either mainstream software or general purpose screen readers. Advancements in graphical user interface design in the early nineties presented great problems of accessibility for visually impaired computer users [43]. However, technical advancements in the design of screen readers greatly improved access to products employing graphical user interfaces such as Microsoft Windows [92], [71]. Yet, not all environments are fully accessible with current screen readers. Screen readers mainly render only textual representations of natural language. Even if many screen readers are able to handle graphical objects such as icons, graphical buttons, and bars, or also complex informative structures such as tables, lists, dialog boxes, and pull-down menus, the description of these objects is almost exclusively textual. Textual representations are often enough to facilitate the construction of a mental image of what is displayed and to interact with the system. Nonetheless, as was mentioned previously, textual descriptions may not always be enough to successfully describe, read, and understand scientific documentation. In order to provide more effective, efficient, and satisfactory techniques to work with complex informative structures (e.g., images, mathematical expressions, and scores), specialized applications are designed.

1.3 Usability of Scientific Documentation through Assistive Technologies

Let us define usability of scientific documentation through assistive tools such as the measure of effectiveness, efficiency, and satisfaction with which blind students can achieve their learning tasks in an educational environment. The factors that affect usability of scientific documentation through tactile and speech assistive technologies are manifold and they are strongly related to usability of text, mathematical expressions, and graphics.

As was mentioned above, text does not present great problems for tactile and speech exploration and editing. Many techniques have been studied and they are currently implemented in mainstream screen readers. Therefore, text usability is generally high through mainstream speech and tactile tools. On the contrary, mathematical expressions present challenging usability issues. Two main factors affect the usability of mathematical expressions: how they can be read and understood, as well as how they can be written and manipulated. In order to comprehend how mathematical expressions can be read and understood through tactile and speech output, it is useful to compare visual reading on the one hand, and tactile and speech reading on the other. A model for visual understanding of mathematical expressions was proposed in 1987 [98]. This model works as follows. A mathematical expression is visually scanned by the reader, whose gaze may rest upon the expression for a while. A mental representation of the surface structure of the expression is formed. This representation is checked for understanding, which involves checking that all symbols are known, and checking that the complexity or length of the expression is manageable. If either of these two tests are failed, the procedure is aborted and a smaller portion of the expression is gazed. Otherwise a syntactic analysis is undertaken in order to come to a tree representation. It is the initial part of this process that poses difficulties for a blind reader [246] as it implies the existence of an external memory (e.g., paper, a blackboard, or a screen) that permanently displays the representation of the mathematical expression. The permanence of the structure to explore and understand relieves the reader of the mental workload in retaining the information to be accessed [231]. Furthermore, the manner in which the information is presented (e.g., whether some portions are underlined or highlighted, how blanks are used to separate blocks, and how the mutual relations among subexpressions are marked) can also help in the process of reading and understanding by sight [158], and the layout of the information can suggest transformations and procedures to perform a certain task. This high-level understanding of the expression is an "overall glance," which can be gotten quickly by sight over the entire expression or over its parts (e.g., subexpressions and building blocks). This ability to get different views allows one to plan and to adapt to the purpose of the reading process. The representation on an external memory can be effective only when its parts can be accessed quickly and accurately. A visual representation on paper or on a screen in conjunction with the control provided by sight allows effective and active understanding. Reading and understanding through speech output is affected by the lack of an external memory. The transient nature of the speech signal does not allow one to have a permanent representation of the structure to read. That causes an inability to get different views of a mathematical expression and to access specific portions directly. As a consequence, the listener tends to be passive [7]. Reading by touch is a more active style of exploration. The external memory is not absent, but it is extremely reduced with respect to the one accessible by sight because it is generally made up of a line of about 40 Braille cells.

Input and editing techniques for mathematical expressions are of paramount importance to achieving high usability of mathematical notations through speech and tactile devices. In particular, they affect the writing speed, which influences the overall efficiency, the ease in carrying out transformations, and the prevention of typing mistakes, which influence both effectiveness and satisfaction. When a sighted person writes a mathematical expression, the flow of information is simultaneously controlled by sight. Not only what is being written can be read by sight, but also a rather large area around the focused piece of information. That allows one to have complete control over the input process and the transformations to be carried out. Consequently, writing mistakes can be immediately located. Touch does not allow simultaneous control over what is being typed in, since the fingers are part of the typing process. However, symbolic patterns can be accurately retrieved by tactile exploration. Hearing may allow synchronous control, but it does not allow the synchronous exploration of already written expressions. Therefore, the design of input and editing techniques that exploit both speech and tactile output is likely to compensate optimally for the lack of sight.

Graphics can be hardly usable through tactile and speech devices. In order to comprehend the reasons that complicate the understanding of graphics in a nonvisual mode, first let us analyze how visual understanding works and which advantages come from a visual exploration of graphical representations. Literature about how sighted people explore and understand graphics suggests some basic features that should be reproduced by any tool for the exploration of nonvisual representations of graphics. A relevant contribution is given by Larkin and Simon [167], who compared the mental computation required in solving problems expounded by diagrams, and problems represented as series of textual elements (e.g., characters, words, and sentences). They found that the mental workload involved in the solution by diagrams is lower than that spent to solve the problem presented through text. Two features of diagram understanding were regarded as the main reasons for the different mental workload: easiness to search and immediacy to recognize. Localization of related parts in diagrammatic representations reduces the need for searching, and consequently it facilitates computation, since symbolic descriptions need not be generated or matched. This means that information represented over a two-dimensional plane can be more efficiently grouped, and searched for meaningful items, than text along a line. As for recognition, diagrammatic representations allow one to understand meaningful shapes immediately, namely, relevant parts can be easily isolated and connected to related diagram components. For example, given the parabola with equation

$$y = -x^2 + 1,$$

it is straightforward to recognize by sight which is the part in the first and second quadrant, when it is displayed, whereas it takes a longer time to get it from a textual description (e.g., through some points in a table). What is natural in the exploration by sight, often becomes difficult in the exploration through speech output and tactile devices. It is due mainly to the transient nature of the speech signal.

2 The LAMBDA System

The LAMBDA (Linear Access to Mathematics for Braille Device and Audio-Synthesis) project is a research project funded by the European Union IST program [232]. It involves working groups with expertise both in assistive technologies and in education for blind students. It aims at developing a system to facilitate writing, reading, and manipulating text and mathematical expressions through Braille displays and speech synthesizers. Further information about the project and free working prototypes can be found at www.lambdaproject.org.

The LAMBDA system is made up of two components: a mathematical Braille code and a multimodal mathematical editor.

In Europe, there is a large variety of Braille codes. Some of them are 6-dot Braille codes and others are 8-dot Braille codes. The 8-dot Braille codes, in particular, are often incomplete and not officially approved by national Braille authorities. Both 6-dot codes and 8-dot codes sometimes lack symbols and rules to represent mathematical expressions. The LAMBDA system aims to provide the user with full Braille output, therefore, in order to overcome these national differences, a set of rules was defined to uniformly describe Braille mathematical expressions in a linear form. Strictly speaking, it was not defined as a unique mathematical Braille code, rather it was developed as a common markup language to represent mathematical expressions linearly so that they can be easily read and understood by means of dot combinations. According to national peculiarities, each symbol is represented through specific configurations of dots. The blind student can read the mathematical expression, not in a national Braille code, after a conversion process, but in a linear form that preserves as far as possible the peculiarities of the national Braille code (e.g., the dots combinations used to represent many mathematical symbols). Each Braille specific tag (e.g., a tag to mark the beginning of a fraction or of a root) is linked to a special glyph so that sighted readers can also understand the linear form of the mathematical expression.

The mathematical editor is the functional component that implements the strategies devised in order to make it easy to read, write, and manipulate text and mathematical expressions by means of speech output and Braille display, in an educational setting. Three groups of working techniques were devised: input, editing, and exploration strategies.

The input of characters and mathematical symbols is achieved through the computer keyboard. Many blind students are used to the key layout of a computer keyboard. Therefore no alternative keyboard, for instance a 8-keys Braille keyboard, was employed. A great deal of mathematical symbols are not present on the computer keyboard, so multiple input strategies were devised and implemented. These strategies take into account the necessity to input quickly, to avoid (as far as possible) input errors, and to reduce the mental workload to remember how to input the needed symbol. For example, it is possible to write the tag opening a subexpression by a mnemonic keyboard shortcut and then input the corresponding closing tag by using the same keyboard shortcut for all open tags. It is also possible to write the name of the tag and have it automatically completed.

As for editing strategies, the mathematical editor implements some general purpose editing operations (e.g., delete the line, delete the word, copy and paste, search and replace) to work with text and mathematical expressions. Other operations are designed especially to facilitate the editing of mathematical expressions. For example, it is possible to delete the content enclosed within a couple of tags, to delete the numerator or the denominator in a fraction, and to substitute corresponding tags for a new couple of tags, which is useful, for example, when a couple of square brackets has to be replaced with a couple of round brackets. These special operations are advantageous both to save time during the editing process and, chiefly, to prevent many mistakes due to the great number of basic operations that would be performed to achieve the same goal. The user can copy the numerator and the denominator into two buffers and then paste them in reverse order in the proper position inside tags for representing fractions. In addition, exploration techniques were designed. Tactile and speech reading usually prevents the blind from getting an overall glance at the mathematical expression and from immediately accessing precise parts inside an expression. Therefore, many operations are provided to move the focus to precise positions in the expression. For example, when the focus is pointing at an open tag, it can be moved automatically to the corresponding separator (e.g., the tag separating numerator and denominator) or to the corresponding closing tag. Further movement operations take into account the mathematical meaning of the structure pointed at by the focus. For example, it is possible to move the focus directly to the denominator or to the numerator, to reach the argument of a function, skipping the function name and some subscripts or superscripts, and so on. One more exploration technique exploits the tag structure of the mathematical expression. For example, the expression

$$-a - [(a - 1) * (a + 1) - 2]$$

can be hierarchically explored as follows:

$$- - a - [] - a - [() * () - 2] - a - [(a - 1) * (a + 1) - 2].$$

In tag structure mode, the mathematical editor enables the user to explore the tag structure of a mathematical expression. The building blocks to hide are determined according to the focus position. The main advantage coming from the structure mode concerns the possibility of getting an overall glance at the linear structure of the expression by reading the building blocks, and, at the same time, reading the symbols actually forming the expression in the unfolded building blocks. Furthermore, it can be used as a table of contents of the expression, so that some precise parts can be immediately accessed without reading every symbol from left to right.

3 Construction of Technical Drawings through a Symbolic Manipulation Software

The drawing strategy through a symbolic language works as follows. A specific language is used to declare the parts of the image, such as shapes, captions, labels, etc. The image is generated by a symbolic manipulation software. The resulting image file can be embossed through a tactile embosser. There is no contextual feedback during the preparation of the image except for the symbolic one. Whether or not the commands to generate the image are carefully grouped, symbolic feedback may be helpful to recognize and search parts of the image being drawn. In the study conducted, Mathematica software [274] was employed for symbolic manipulation and the Tiger graphical embosser was used to print tactile images. Three reasons led to the choice of Mathematica:

1. The language used by Mathematica has rather simple and uniform syntactic rules.
2. It has a great deal of built-in functions and it is easy to extend them to create specific packages.
3. It embeds many functions to work with XML structures. That allows, for example, one to place Mathematica objects, such as images, inside XHTML pages for further exploration with other techniques.

Tiger embosser was chosen because it is able to reach high dot resolution. In order to obtain high-quality tactile images, a 25.4 DPI (dots per inch) resolution is necessary [236]. Tiger embossers are able to print with a 20 DPI resolution, which is near to the best one. Furthermore, automatic conversion of colored and shaded images to tactile graphics with variable height dots can be achieved, so that relevant shapes

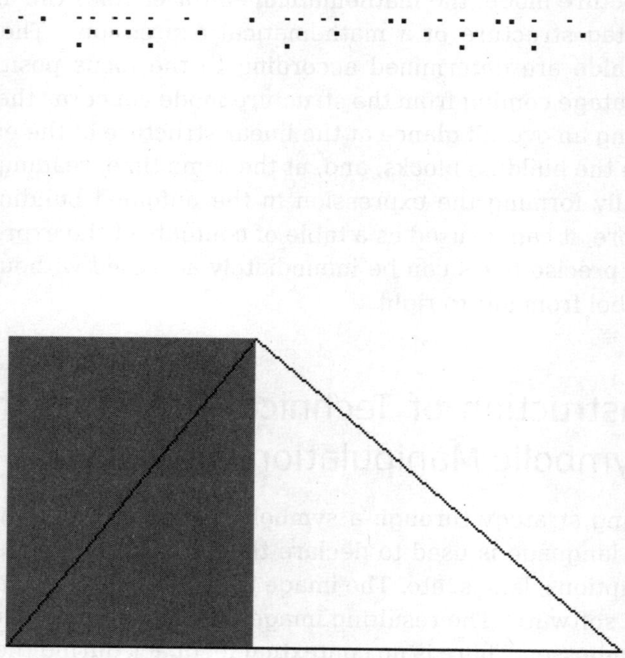

Figure 1. Triangle and rectangle with Braille title.

can be perceived higher than less important ones. The graphical primitives in Mathematica were used to define a set of graphical primitives oriented to the preparation of tactile images. These specific primitive functions are meant to reduce syntactic complexity and to hide the setting of specific parameters necessary for a good tactile rendering (e.g., the definition of Braille labels). The advantages of this technique for tactile drawing preparation are described through a use case.

Let us assume that a user wants to prepare a tactile image with a triangle intersected with a rectangle having a diagonal coincident with a side of the triangle. That can be achieved as follows:

```
(* The title is set *)
title = Text["triangle and rectangle", {7,30} ];
(* Triangle PQR *)
P={5,0}; Q={10,0}; R={7,10};  trianglePQR=Triangle[ P,Q,R ];
(* Rectangle *)
rectangle = Bright[ Rectangle[P,R] ];
(* they are in the same figure *)
figure = {title, rectangle, trianglePQR};
(* The figure is embossed by Tiger *)
TactilePrint[figure];
```

The graphical result is shown in Figure 1. Let us analyze what Mathematica actually executes, namely, which operations are hidden by the primitives for tactile rendering.

```
title = Graphics[Text["triangle and circle", {7,20} ]];
P={5,0}; Q={10,0}; R={7,10};  trianglePQR=Graphics[Line[ {P,Q,R,P} ]];
rectangle = Graphics[ {GrayLevel[0.5],Rectangle[ P,R ]} ];
options = {TextStyle->{ FontFamily->"Braillekiama", FontSize->24}, AspectRatio->1};
figure=Show[ {title, rectangle, trianglePQR}, options];
Export["temporaryimage.bmp",figure];
Run["print_to_tiger","temporaryimage.bmp"];
```

Some observations can be made:

- The primitives for graphical objects (e.g., triangle, rectangle) allow one to recognize basic shapes even when they are not drawn. Instead, many difficulties are met to recognize mutual relations among basic shapes.
- Basic components of the image can be described by primitive functions and assigned to variables. These variables can be grouped in lists (e.g., figure) that describe the actual image. Each component can be reused in other images. Moreover, more complex components (e.g., an inclined plane to describe certain experiments in physics) can be created and reused through translation and rotation directives.

- Textual labels are embossed in Braille. Braille characters are obtained applying a Braille font. In this use case the Braillekiama True Type font was applied. It is a 6-dots Braille font.
- The size of the image can be set in the options. Any way, automatic adaptations are performed. Images too small cannot be properly perceived by touch.
- The rectangle is printed as bright. This means that when the figure is printed, the Tiger embosser fills the rectangle with dots lower than those making the triangle. Thus, both the polygons are discernible by touch.

4 Conclusions

The design of educational tools to improve the working strategies employed by blind students in reading, writing, and manipulating scientific documentation has been studied. A mathematical editor, developed through the LAMBDA project, provides the core of a framework that can group together the different modules implementing specific working strategies (e.g., those concerning tactile diagram construction). Therefore, all of these tools will be completely useful once they are collected in a uniform working environment, accessible through the user interface provided by the mathematical editor. Further extensions are planned, especially concerning evaluation and the tools to exchange data with mainstream applications.

On the Conversion between Content MathML and OpenMath

C.M. So and S.M. Watt

Content MathML and OpenMath are two XML formats for semantic markup of mathematical expressions. Although efforts have been made to align their definitions, enough differences remain to make translation between them nontrivial.

In this chapter, we present a technical discussion of the differences between Content MathML and OpenMath, and two strategies for bidirectional translation between them. The first approach is to map between pre-defined Content MathML elements and standard OpenMath definitions, making any necessary adaptations for concepts that do not exactly correspond. For the second method, we describe a mapping using the extension mechanisms of each language, wherein all references to standard concepts are replaced by equivalent external references.

We have implemented these translation schemes (both approaches and both directions) as XSLT style sheets. These implementations have served both to test our ideas and as components in an architecture for mathematical web services.

1 Introduction

Mathematical software packages, including computer algebra systems and theorem provers, define their own system-dependent syntax for commands and expressions. This is appropriate under the assumption that a person uses only a small number of such systems. The drawback of package-specific notation for mathematical content is that it makes data exchange between programs difficult and it limits a person's ability to take advantage of the strengths of unfamiliar systems.

OpenMath [73,80,257] and Content MathML [17] were both devised to provide system-independent, nonproprietary formats for the semantics of mathematical expressions. They are complementary standards with different emphases. For example, OpenMath currently does not define how its expressions should be displayed, and can use MathML for that purpose. Conversely, Content MathML may be extended using OpenMath annotations to express semantic concepts that it does not natively provide. The relationship between OpenMath and MathML has been discussed elsewhere [53,55,193].

There are two implications to the fact that the standards are complementary. First, since each standard may rely to some extent on the other, it is clear that the languages are needed simultaneously for the general management of mathematical knowledge (on the web or otherwise). Hence, translations between the languages are necessary. However, the codependence between the languages also implies that possibly they are different on a deeper level, that is, sufficiently mismatched so as to complicate such translations. Indeed, as we will describe, there are differences that prevent the simple translation of certain expressions from conserving semantics.

Our primary aim in this chapter is to give a detailed analysis of translation between the Content MathML and OpenMath formats. In doing so, we uncover certain discrepancies that may preclude the full interoperability of the languages, and which therefore should be taken into consideration in future versions of these standards.

We describe two strategies for translation between Content MathML and OpenMath, and we demonstrate how these mismatches may be addressed in both cases. The first strategy is to convert between predefined Content MathML elements and OpenMath's MathML Content Dictionaries, preserving the use of Content MathML's built-in semantics if possible. This translation also handles the primitives of both formats. The second strategy maps all semantics of the source into expressions using the extension mechanism of the target language. This open-ended strategy uses only the low-level expression forming primitives of OpenMath and Content MathML, and uses external definitions for all mathematical content. Both strategies match the basic elements of the standards.

To summarize, the principle contributions of this work are:

- *A strategy for high-level conversion between OpenMath and Content MathML.* We identify mismatches between Content MathML and OpenMath, including shortcomings of the OpenMath "MathML" Content Dictionary, intended to support Content MathML.

- *A strategy for uniform conversion between OpenMath and MathML.* The target language, either OpenMath or Content MathML, is used as a "carrier" syntax, and translations uniformly use the extension mechanisms.

The remainder of this chapter is organized as follows. Section 2 introduces the Content MathML and OpenMath standards. Section 3 presents the mapping between the semantics of Content MathML and OpenMath's MathML CD group. It also presents the discrepancies between Content MathML and OpenMath. Section 4 describes how we can always use external definitions in Content MathML when matching semantics are not available. This translation strategy helps us to overcome the cases in which OpenMath's semantics do not have Content MathML equivalent. Section 5 describes our implementation of a translator based on these ideas, and the chapter ends with our conclusions.

2 OpenMath and Content MathML

The OpenMath and Content MathML standards both provide facilities to manage mathematical knowledge. Both formats serve to encode the semantics of mathematical expressions. Neither format is used to specify mathematical notation. Rather, it is expected that $\TeX$ or Presentation MathML can be used to represent expression notation, which is beyond the scope of this chapter. Before contrasting the two languages, we highlight their key characteristics individually.

2.1 OpenMath

OpenMath is a system-independent XML [44] format to encode the semantics of mathematical expressions. It can be used as input and output for scientific computation, as well as the format for mathematical expressions in documents. This standard includes a set of predefined elements describing some elementary concepts such as variables, floating point numbers, and integers. No mathematical function is predefined. All semantics of the mathematical functions are defined in collections called *Content Dictionaries* (CDs). These CDs may be defined by a single application, by an agreement between two parties, or by community consensus.

Example 1. $\int x^2\, dx$ in OpenMath.

```
<OMOBJ xmlns="http://www.openmath.org/">
  <OMA>
    <OMS cd="calculus1" name="int"/>
    <OMBVAR> <OMV name="x"/> </OMBVAR>
    <OMA>
      <OMS cd="arith1" name="power"/>
      <OMV name="x"/> <OMI> 2 </OMI>
    </OMA>
  </OMA>
</OMOBJ>
```

Extending OpenMath. OpenMath is extensible by design. The semantics of mathematical functions are defined using Content Dictionaries, which may, in principle, be contributed by anyone. This removes the need to refer to semantics expressed in languages other than OpenMath. Existing, frequently used mathematical functions and constants are predefined in official "standard" OpenMath CDs.

The standard CDs and CD groups of OpenMath are organized so that related functions and constants are usually placed in the same CD. Related CDs then belong to a CD group. The OpenMath Society has already approved a certain number of CDs and CD groups. In the next section, we examine one of the OpenMath Society's official CDs, the MathML CD group, shown in Figure 1.

2.2 Content MathML

MathML [17] is a system-independent format to encode mathematical expressions for web documents. Different types of MathML markup are available to accommodate different applications of mathematics. *Presentation markup* provides a format for specifying the notation of a mathematical expression. *Content markup* provides a format for specifying the semantics and the structure of mathematical expressions without implying any actual notation being used. This type of markup allows applications to specify the notation, and process the semantics of the expression separately. *Mixed markup* associates the two types of markup by bundling together, at some level of resolution, the notation and the semantics of a mathematical expression. This type of markup allows a variety of applications to process the expression, using either the semantics or the notation.

Since we aim to convert between OpenMath and MathML, we are concerned only with the Content markup aspect of MathML. Currently, Content MathML consists of a base set of elements covering subjects in elementary mathematics. The subject areas covered to some extent

CD name	Description
alg1	Some basic algebraic concepts
arith1	Common arithmetic functions
bigfloat1	Representation of floating point numbers
calculus1	Calculus operations
complex1	Operations and constructors of complex numbers
fns1	Constructors and functions for functions
integer1	Basic integer functions
interval1	Discrete and continuous 1-dimensional intervals
linalg1	Operations on matrices
linalg2	Matrices and vectors in a row oriented fashion
limit1	Basic notion of the limits of unary functions
list1	List constructors
logic1	Basic logic functions and constants
mathmltypes	Types and constructs handled by MathML
minmax1	Minimum and maximum of a set
multiset1	Basic multiset theory
nums1	Common numerical constants and constructors
piece1	Operators for piece-wise defined expressions
quant1	Basic universal and existential quantifiers
relation1	Common arithmetic operations
setname1	Common sets in mathematics
rounding1	Basic rounding concepts
set1	Functions and constructors for basic set theory
s_data1	Basic statistical functions used on sample data
s_dist1	Basic statistical functions used on random variables
transc1	Transcendental functions
veccalc1	Functions for vector calculus
altenc	Alternative encodings

Figure 1. OpenMath's MathML CD group.

by Content MathML include arithmetic, algebra, logic, relations, calculus, set theory, sequences and series, elementary classical functions, statistics, and linear algebra.

Example 2. $\int x^2\, dx$ in Content MathML.

```
<math xmlns="http://www.w3.org/Math/">
  <apply>
    <int/>
    <bvar> <ci> x </ci> </bvar>
    <apply> <power/> <ci>x</ci> <cn>2</cn> </apply>
  </apply>
</math>
```

Extending Content MathML. Content MathML's base set of elements cannot natively handle more than a fraction of the vast set of mathematical concepts. To overcome this problem, Content MathML provides an extension mechanism allowing one to redefine Content MathML elements or to introduce new mathematical semantics. The following examples illustrate the use of these extension mechanisms in Content MathML.

Example 3. Content MathML predefines `<minus>` as a unary or a binary arithmetic operator. If we would like to use this element as a n-ary operator, we have to override its semantics. This is accomplished by referring to an external URL. The "definitionURL" attribute gives a location defining the new semantics. The following Content MathML markup illustrates how one might encode the expression $1-2-3-4-5$:

```
<apply>
  <minus definitionURL=
        "http://www.orcca.on.ca/example/MathML/n-ary#new_minus"/>
  <cn>1</cn> <cn>2</cn> <cn>3</cn> <cn>4</cn> <cn>5</cn>
</apply>
```

A second example shows how to introduce completely new mathematical functions into Content MathML. The `<csymbol>` element is used for this purpose.

Example 4. Bessel functions of the first kind are not defined in Content MathML. To introduce these functions, we need to specify a URL giving the semantics as an attribute within a `<csymbol>` element. The following Content MathML markup illustrates how to define and use the function:

```
<apply>
   <csymbol definitionURL=
     "http://www.orcca.on.ca/example/OpenMath/SpecialFns/bessel#BesselJ"
     >BesselJ</csymbol>
   <cn> 0 </cn>
   <cn> 0 </cn>
</apply>
```

3 Conversion Using Internal Semantics

On first examination, Content MathML and OpenMath appear quite similar. One-to-one correspondences can be drawn between most of the basic elements and between some mathematical elements as well. This is shown in the example in Figure 2. For example, `<apply>` maps to

```
<math>
  <apply>
    <equivalent/>
    <apply>
      <not/>
      <apply>
        <and/>
        <ci> A </ci>
        <ci> B </ci>
      </apply>
    </apply>
    <apply>
      <and/>
      <apply>
        <not/>
        <ci> A </ci>
      </apply>
      <apply>
        <not/>
        <ci> B </ci>
      </apply>
    </apply>
  </apply>
<math>
```

```
<OMOBJ>
  <OMA>
    <OMS cd="logic1" name="equivalent"/>
    <OMA>
      <OMS cd="logic1" name="not"/>
      <OMA>
        <OMS cd="logic1" name="and"/>
        <OMV name="A"/>
        <OMV name="B"/>
      </OMA>
    </OMA>
    <OMA>
      <OMS cd="logic1" name="and"/>
      <OMA>
        <OMS cd="logic1" name="not"/>
        <OMV name="A"/>
      </OMA>
      <OMA>
        <OMS cd="logic1" name="not"/>
        <OMV name="B"/>
      </OMA>
    <OMA>
  </OMA>
</OMOBJ>
```

Figure 2. Content MathML (left) and OpenMath (right) markup of $\neg(A \wedge B) \equiv (\neg A \vee \neg B)$.

<OMA>, <ci> maps to <OMV>, and the <OMS>s map to their corresponding MathML mathematical functions.

The OpenMath MathML CD group is designed to provide MathML compatibility. It is intended to enable simple translation, mapping between functions described by this CD group and Content MathML elements. Such a translation is reported in [239].

Translation between these two formats is not always so simple, however. Although OpenMath's MathML CD group is designed to be compatible with Content MathML, some mismatches between OpenMath and Content MathML exist. Specifically, some elements present in one language are not present in the other, and some elements present in both languages have slightly different meanings or usage. A robust translator must deal with differences in the meanings of the predefined mathematical functions as well as expressions that make use of the extension mechanisms in both languages.

3.1 Dealing with the Differences

There are three ways to handle differences between Content MathML and OpenMath within a translation: If usage is different between two matching elements, a special transformation can be performed. In cases where there is more than one possible match, heuristics can be derived for eliminating inappropriate choices. Finally, failing these two options, we can extend Content MathML by making external reference to OpenMath's semantics.

Special transformation. Although there is some overlap between the development communities of OpenMath and Content MathML, differences in usage can be found in matching language elements. In such cases, we can construct the mathematical object based on the information within the existing markup. This solution can be used in both directions of the translation. Example 5 illustrates this approach.

Example 5. Intervals are expressed differently in Content MathML and OpenMath. This example shows the interval $(0, \pi)$ in Content MathML (left) and OpenMath (right). Knowing the difference between the two formats, we simply perform a special transformation to reconstruct the appropriate elements for either target.

```
<interval closure="open">
  <cn type="integer">0</cn>
  <pi/>
</interval>
```

```
<OMA>
  <OMS cd="interval1" name="interval_oo"/>
  <OMS cd="alg1" name="zero"/>
  <OMS cd="nums1" name="pi"/>
</OMA>
```

Heuristics. Many of Content MathML's function elements have more than one usage. They can take various numbers or types of arguments. OpenMath's <OMS>s do not tend to have alternative usages. To determine the proper OpenMath <OMS> of a Content MathML function, it is necessary to examine the arguments.

Example 6. Content MathML's `<int/>` is used for both definite and indefinite integration. It has two corresponding OpenMath <OMS> attributes: `defint` from the `calculus1` CD specifies a definite integral and `int` from the same CD specifies an indefinite integral. To determine the correct translation to OpenMath, it is necessary to inspect the arguments of the function.

Heuristics can be used to deal with usage differences, as shown in the following example.

Example 7. Content MathML's `<partialdiff>` (below left) accepts bound variables as a list of indexes or series of actual variable names. The OpenMath equivalent of `<partialdiff>` (below right) does not recognize the variables as a series of variable names. To deal with this difference, we can assume x maps to index 1 and so on.

```
<apply>                          <OMA>
  <partialdiff/>                   <OMS cd="calculus1" name="partialdiff"/>
  <bvar>                           <OMA>
    <ci> x </ci>                     <OMS cd="list1" name="list"/>
  </bvar>                            <OMI> 1 </OMI>
  <bvar>                             <OMI> 2 </OMI>
    <ci> y </ci>                   </OMA>
  </bvar>                          <OMA>
  <apply>                            <OMV name="f"/>
    <ci> f </ci>                     <OMV name="x"/>
    <ci> x </ci>                     <OMV name="y"/>
    <ci> y </ci>                   </OMA>
  </apply>                       </OMA>
</apply>
```

External semantics. If neither special transformations nor heuristics are sufficient to provide a faithful translation, one may use the Content MathML extension mechanism (Section 2.2) to refer to an OpenMath definition. To convert from OpenMath to Content MathML, a `definitionURL` attribute is constructed, using the `CD` and `name` attributes from the OpenMath `<OMS>` element, and added to the target Content MathML `<csymbol>` element. To do the reverse transformation, the `CD` and `name` attributes can be extracted from the `definitionURL` attribute.

Example 8. Content MathML does not have an equivalent of the definitions in the `bigfloat1` CD.

```
<apply>
  <csymbol encoding="OpenMath"     <OMA>
    definitionURL=                   <OMS cd="bigfloat1" name="bigfloatprec"/>
    "http://www.openmath.org/(no break)  <OMV name="m"/>
      cd/bigfloat1.ocd#bigfloatprec"   <OMV name="r"/>
  <ci> m </ci>                       <OMV name="e"/>
  <ci> r </ci>                     </OMA>
  <ci> e </ci>
</apply>
```

3.2 Examples of Mismatches

We now discuss how these strategies can be used in dealing with the differences between Content MathML and OpenMath. The examples also illustrate the strategies' limitations. Note that the strategies do not provide a one-to-one mapping between the two formats. Instead, they help preserve, as much as possible, semantics under translation. For some of the mismatches, none of the strategies are applicable. This situation is discussed in Section 4.

Type information. In OpenMath, any mathematical object, including compound objects, can be annotated with type information. Types are

```
                                  <OMATTR>
                                    <OMATP>
                                      <OMS cd="mathmltypes" name="type"/>
<ci type="integer">n</ci>             <OMS cd="mathmltypes" name="integer_type"/>
                                    </OMATP>
                                    <OMV name="n"/>
                                  <OMATTR>
```

Figure 3. Content MathML (left) and OpenMath (right) giving the symbol n integer type.

defined in CDs, and mathematical objects are annotated using <OMATTR> and <OMATP>, which are the "attribute constructor" and "attribute pair" elements. In Content MathML, type information can be specified directly only for numbers and identifiers, using the type attribute within the elements <cn> and <ci>, as shown in Figure 3. Although it would be possible to provide MathML annotations, using <semantics> and <annotation> or <annotationXML>, this does not place the information into Content MathML. The simple heuristic to apply in the translation of OpenMath type-attributed objects is: if the type-attributed OpenMath object is an integer or variable, then a special transformation can be applied; otherwise, employ the Content MathML extension mechanism.

Numbers with compound structure. OpenMath uses CDs to define numbers with compound structure, such as rational numbers (Figure 4). A number may be given in parts, which, in general, may themselves be compound objects. Content MathML, however, cannot represent numbers as a composition of compound mathematical objects. Rather, in Content MathML, a number must have components that are simple tokens given as text. The interpretation of the components is specified by the type attribute, which must be one of real, integer, rational, complex-cartesian, complex-polar or constant. If, as shown in Figure 4, an OpenMath number can be represented in one of the forms allowed by Content MathML <cn>, a special transformation can be applied. Otherwise, external semantics may be used or a composite

```
<cn type="rational">              <OMA>
  2                                 <OMS cd="nums1" name="rational"/>
  <sep/>                            <OMI> 2 </OMI>
  3                                 <OMI> 3 </OMI>
</cn>                             </OMA>
```

Figure 4. Rational number by Content MathML (left) and OpenMath (right).

```
<apply>                     <OMA>
  <int/>                      <OMS cd="calculus1" name="int"/>
  <bvar>                      <OMBIND>
    <ci> x </ci>                <OMBVAR> <OMV name="x"/> </OMBVAR>
  </bvar>                       <OMA>
  <apply>                         <OMS cd="fns1" name="lambda"/>
    <sin/>                        <OMA>
    <ci> x </ci>                    <OMS cd="transc1" name="sin"/>
  </apply>                          <OMV name="x"/>
</apply>                          </OMA>
                                </OMA>
                              </OMBIND>
                            </OMA>
```

Figure 5. $\int \sin(x)\, dx$ in Content MathML (left) and OpenMath (right).

Content MathML may be constructed (i.e., as an expression, rather than a single <cn>).

Use of lambda bindings. In OpenMath, lambda bindings can be used to represent functions, as shown in Figure 5. For example, the `calculus1` CD operators `diff` and `int` can take a function whose arguments are specified with lambda bindings. In Content MathML, however, bound variables are not specified as part of the function, but rather as arguments to the operators. So, in translating from OpenMath to Content MathML, if a function expression does not give an explicit lambda binding, then there is no way to determine the bound variables. If the argument of such functions in OpenMath is not a lambda binding, we may use heuristics to guess the bound variable when translating from OpenMath to Content MathML. In the reverse translation, we need to reconstruct a lambda binding from the function application with the bound variables.

Miscellaneous integer functions. There is no direct Content MathML equivalent of OpenMath's `trunc` and `round` from the `rounding1` CD, and `abs` from the `arith1` CD. External references or special transformations constructing equivalent mathematical functions may be used.

We have seen that while the definitions of OpenMath and Content MathML allow many expressions to be translated precisely, there are several technical points on which the formats do not agree. Provided the application can be made to avoid these aspects, it is possible to have a natural translation between the corresponding concepts provided by the two standards. If a more robust solution is required, then another approach is needed, as described in the next section.

4 Conversion Using External Semantics

For a completely general solution to the problem of translation between OpenMath and Content MathML, it is necessary to forgo the translation between high-level concepts that sometimes do not match exactly. Instead, we must rely on mechanisms that give correct translation under all circumstances. In our approach we have relied on the general extension mechanisms of OpenMath and MathML. We therefore have a general, uniform way to convert between Content MathML and OpenMath.

In this setting, the conversion from Content MathML to OpenMath is similar to that described in Section 3. The only difference is that new CDs, precisely capturing the Content MathML semantics, may be used.

In contrast, the conversion from OpenMath to Content MathML is now quite different. We use a simple subset of Content MathML elements that correspond to the OpenMath expression-forming elements. These include `<ci>`, `<cn>`, `<csymbol>`, `<lambda>`, `<bvar>` and `<apply>`. Then *all* OpenMath symbols are translated to `<csymbol>` elements with references to the canonical OpenMath URLs. Effectively, we embed OpenMath in Content MathML syntax. In other words, no attempt is made to generate "native" Content MathML.

An example of such a conversion is given in Figure 6. The basic OpenMath constructs, such as the apply construct, `<OMA>`, are mapped to the corresponding Content MathML constructs. The symbol construct, `<OMS>`, maps to the MathML `<csymbol>` element, with an attribute containing the URL for the OpenMath semantics.

This strategy not only provides a solution to translating OpenMath elements for which no corresponding Content MathML construct ex-

```
<math>
  <apply>
    <cysmbol definitionURL=
     "http://www.openmath.org/(no break)
      cd/mathml/logic1#not"/>
    <apply>
      <cysmbol definitionURL=
       "http://www.openmath.org/(no break)
        cd/mathml/logic1#and"/>
      <ci> A </ci>
      <ci> B </ci>
    </apply>
  </apply>
<math>
```

```
<OMOBJ>
  <OMA>
    <OMS cd="logic1" name="not"/>
    <OMA>
      <OMS cd="logic1" name="and"/>
      <OMV name="A"/>
      <OMV name="B"/>
    </OMA>
  </OMA>
</OMOBJ>
```

Figure 6. "Simplified" Content MathML (left) and OpenMath (right) markup of $\neg(A \wedge B)$.

ists, but also illustrates how OpenMath can be used to complement Content MathML's limited facilities for expressing the semantics of a wider range of mathematical subjects.

5 Implementation

We have implemented a translator based on the strategies discussed in Sections 3 and 4. The translator was implemented as a set of XSLT [65] style sheets. The style sheets are tested using `xt` [277], one of the implementations of XSLT. XSLT was chosen to implement the translator because it is designed specifically to transform XML expression trees. Another advantage is that many implementations of XSLT are free and widely available. The style sheets are available at the website of our laboratory: [url:154].

There has been at least one other implementation of a Content MathML / OpenMath translator [239]. This previous work discusses only the correspondence between the built-in semantics of the formats and is based in a REDUCE environment. In contrast, our open-ended conversion strategy does not restrict our attention to built-in semantics, and we have employed more widely available XML technology for our implementation.

It should be noted that our work is based on OpenMath version 1.0 and MathML 2.0. Since the development of our translator, both of the standards have evolved. At the time of writing, OpenMath 2.0 has been released and discussions for MathML 3.0 are underway. Although minor revisions will be made to the respective standards, our translation strategies can still be applied and the mismatches that we have identified here still exist.

6 Conclusion

OpenMath and Content MathML are two standards to encode the semantics of mathematical expressions. There are many similarities and differences between them.

We have described two translation strategies to convert between these two languages: The first strategy maps between the corresponding elements of the standards. It exposes a number of differences between OpenMath and Content MathML that must be dealt with. The second strategy uses the low-level structure of the formats to give precise embeddings of each within the other.

We have implemented these translation strategies using XSLT. The implementations have served two purposes: first as a proof of concept, to verify the validity of the approaches, and second, to fill a practical need, providing translator blocks in an architecture for mathematical web services [189].

XML-Based Format for Descriptions of Geometrical Constructions and Proofs

P. Quaresma, P. Janičić, J. Tomašević, M. V.-Janičić, and D. Tošić

With a large number of tools focusing on visualizing geometrical constructions or on proving properties of constructed objects (or both), there is an emerging need for linking them, and making them and their corpora widely usable. A common setting that links these tools would be important in the field of geometrical constructions and in their role in education. In the following text, we propose a common XML-based interchange format for descriptions of geometrical constructions and proofs. We also present an XML library providing support for dynamic geometry software and automatic theorem provers, and their integration into our web-based GeoThms system.

1 Introduction

Dynamic geometry software (DGS), such as Cinderella [url:87], Geometer's Sketchpad, [url:123], and Cabri [url:76], visualize geometric objects and link the formal, axiomatic nature of geometry, most often Euclidean, with its Cartesian models and corresponding illustrations. The common experience tells us that dynamic geometry tools significantly help students in acquiring knowledge about geometric objects and, more generally, in acquiring mathematical rigor.

In many DGSs, a geometric construction is specified using, explicitly, a formal language. In others, the construction is made interactively, by clicking specific buttons and/or icons, but behind this approach there is also a formal geometrical language, although usually hidden from the user. All these languages share many primitive commands (related to geometrical constructions), but there are also differences in the set of supported commands, and they follow different syntax rules.

Besides DGSs, there are automated theorem provers (ATPs) specialized for geometrical constructions. Some of them aim at producing traditional, human readable geometrical proofs [45,145,192,206,279].

With a large number of tools focusing on visualizing geometrical constructions or on proving properties of constructed objects (or both), there is an emerging need for linking them and making widely usable constructions and proofs generated with different tools. This would help in the progress of the field of geometrical constructions, including their role in education.

We believe that descriptions of geometrical constructions and geometrical proofs should be put into the XML framework, by defining a *normal form*, linked to different formats. In this paper, we describe an XML-based system built on such an XML-based format. These are some of the most important motivating arguments for using XML in storing descriptions of geometrical constructions and proofs, and as an interchange format:

- Instead of raw, plain text representation, geometrical constructions will be stored in strictly structured files; these files will be easy to parse, process, and convert into different forms and formats.
- Input/output tasks will be supported by generic, external tools and different geometry tools will communicate easily.
- A growing corpora of geometrical constructions will be unified and accessible to users of different geometry tools.
- There will be easier communication and exchange of material with the rest of the mathematical and computer science community.
- There is a wide and growing support for XML.
- Different sorts of presentation (text form, LaTeX form, HTML) will be easily enabled.
- There will be strict content validation of documents with respect to given restrictions.

We have implemented converters for two DGSs, confirming, in this way, that the proposed XML format can serve its main purpose. We have also developed XML support for automatically generated proofs of constructive geometrical theorems. These tools, together with rendering tools (tools for visual presentation of XML files) were incorporated into our GeoThms framework.

2 Background

In this section, we give some basic background information about geometrical constructions, XML, and our GeoThms framework that links DGSs, ATPs, and a repository of geometry problems.

2.1 Geometrical Constructions

For hundreds, or even thousands of years, geometric construction problems have been one of the most attractive parts of geometry and mathematics. A geometric construction is a sequence of specific, primitive construction steps. These primitive construction steps (also called *elementary constructions*) are based on using a ruler (or a straightedge[1]) and a compass, and they are:

- Construction (with ruler) of a line such that two given points belong to it.
- Construction (with ruler) of a segment connecting two points.
- Construction (with compass) of a circle such that its center is one given point and such that the second given point belongs to it.
- Construction of a point that is the intersection of two lines (if such a point exists).
- Construction of intersections between a given line and a given circle (if such points exist).

By using the set of primitive constructions, one can define more complex constructions (e.g., the construction of a right angle and the construction of the midpoint of a line segment).

The abstract (i.e., formal, axiomatic) natures of geometric objects must be distinguished from their usual interpretations. A geometric construction is a procedure consisting of abstract steps and it is not a picture, but for each construction there exists its counterpart in the standard Cartesian model.

[1]The term "straightedge" is sometimes used instead of "ruler" in order to emphasize that there are no markings which could be used to make measurements.

2.2 XML

The *eXtensible Markup Language* (XML) is a simple, very flexible text format for data structuring using tags, inspired by SGML (ISO 8879). Originally designed to meet the challenges of large-scale electronic publishing, XML is also playing an increasingly important role in the exchange of a wide variety of data on the web and elsewhere [url:188]. It is called "extensible" because it is not a fixed format like HTML (a single, predefined markup language); instead the tags indicate the semantic structure of the data, rather than (only) its layout in a browser. XML is actually a "metalanguage"— a language for describing other languages, which lets one design his/her own customized markup languages for limitless different types of documents. XML provides a structured way of transmitting information between programs and systems. It is intended to make it easy to define document types, to write and maintain documents, and to share them across the Internet.

However, XML is not just for webpages: it can be used to store any kind of structured information, and to enclose or encapsulate information in order to pass it between different computing systems. An XML document can carry both presentation (i.e., plausible visualization) and content information. XML is a project of the World Wide Web Consortium (W3C) and is a public format—it is not a proprietary development of any company. Almost all browsers that are currently in use support XML natively.

Data type definitions (DTDs) provide a formal specification of the constraints on the structure of data presented in XML form. A DTD is given as a formal description in XML declaration syntax. It sets out what are the names to be used for the different types of elements, where they may occur, and how they all fit together. This formal description enables automatic verification ("validation") of whether a document meets the given syntactical restrictions. This way, groups sharing data of a similar sort can agree on their XML representation and corresponding DTDs.

Extensible style sheet language transformation (XSLT) is a document-processing language that is used to transform the input XML documents to output files. An XSLT style sheet declares a set of rules (templates) for an XSLT processor to use when interpreting the contents of an input XML document. These rules tell the XSLT processor how that data should be presented: as an XML document, as an HTML document, as plain text, or in some other form.

Scalable Vector Graphics (SVG) is a language, based on XML, for describing two-dimensional graphics and graphical applications. As for other XML applications, there is a W3C recommendation for SVG [url:181].

2.3 GeoThms Framework

GeoThms [url:29] is a Web workbench in the field of constructive problems in Euclidean geometry. It is a framework that links dynamic geometry software, geometry automatic theorem provers, and a repository of geometry problems (geoDB), providing a common web interface for all these tools. Its tight integration of dynamic geometry tools and automatic theorem provers and its repository of theorems, figures and proofs, gives the user the possibility to easily browse through the list of geometric problems, their statements, illustrations and proofs. Currently, there are the following tools integrated in GeoThms:

- GCLC[2] [144] and Eukleides[3] [196, 207] are two DGSs; they both use (similar) geometry drawing languages in which producing mathematical illustrations is based on "describing figures," rather than on "drawing figures." These descriptions directly reflect the meaning of mathematical objects to be presented, and are easily understandable to mathematicians. Both tools have graphical user interfaces and the ability to produce LaTeX files with illustrations for geometrical constructions.
- GCLCprover [145, 206] and CoqAreaMethod [192] are two ATPs based on the area method [62, 63]. GCLCprover is a theorem prover that allows formal deductive reasoning about objects constructed with the help of DGSs. It produces proofs that are human-readable, and with a clear justification for every proof step. GCLCprover is tightly integrated with the GCLC, which means that one can use the prover to reason about a GCLC construction, without changing and adapting it for the deduction process. Users only need to add a statement that they want to prove. The geometrical constructions made within GCLC are internally transformed into primitive constructions of the area method, and in certain cases, some auxiliary points are introduced. With support

[2] The GCLC package is freely available from www.matf.bg.ac.yu/˜janicic/gclc/. There are versions for Windows and Linux.

[3] Eukleides is available from [url:113]. The first author of this paper is responsible for the Portuguese version of Eukleides: EukleidesPT is available from [url:27].

of our XML library, it is also possible to reason about the Eukleides constructions. CoqAreaMethod implements the area method within the Coq proof assistant implements [100].

- The geoDB database gives support to the other tools, keeping the information, and allowing for its fast retrieving whenever necessary. Constructions are described and stored in XML form. Figures are generated from the XML files, by DGSs, and stored in suitable formats (JPEG and SVG). Conjectures are described and stored in a form that extends geometric specifications. The specifications of conjectures are used via converters by ATPs. Proofs are generated by ATPs and stored in suitable formats (PDF and XML in compressed form).

3 Overall Architecture

In this section, we provide some motivating arguments for introducing an XML-based format in representing geometrical constructions and geometrical proofs. Also, we propose the architecture of a system based on these motivations and ideas (the actual implementation of our system is described in the next section).

3.1 Representation of Construction Descriptions

All dynamic geometry tools use some formal languages for describing geometrical objects (either a hidden, underlying language or a user-oriented language). Consider, for instance, two equivalent descriptions (in GCLC language and in Eukleides language) of the same construction given in Figure 1. GCLC language and Eukleides language were developed/defined independently by independent authors. Corresponding descriptions in languages of many other geometry tools are similar. The reason for this is that all these tools describe the elementary constructions by ruler and compass (see Section 2.1) and deal with similar additional requests for drawing and labeling geometrical figures. So, all of these languages are very similar, but still different (due to different main purposes, different authors, different implementations, etc.).

In order to enable communication between these tools and conversion of files between different formats, it is good to have a single target format—a format that defines a common normal form for different tools. We propose one such format, within a general XML specification.

`dim 80 80`	`frame(0,0,8,8)`
`point A 10 30`	`A = point(1,3)`
`point B 60 10`	`B = point(6,1)`
`point C 50 70`	`C = point(5,7)`
`med a B C`	`a = bisector(segment(B,C))`
`med b A C`	`b = bisector(segment(A,C))`
`med c B A`	`c = bisector(segment(B,A))`
`intersec O_1 a b`	`O1 = intersection(a,b)`
`intersec O_2 a c`	`O2 = intersection(a,c)`
`drawline a`	`draw(a)`
`drawline b`	`draw(b)`
`drawline c`	`draw(c)`
`drawsegment A B`	`draw(segment(A,B))`
`drawsegment A C`	`draw(segment(A,C))`
`drawsegment B C`	`draw(segment(B,C))`
`cmark_lb A`	`draw(A); label(A,-90:)`
`cmark_b B`	`draw(B); label(B,-90:)`
`cmark_t C`	`draw(C); label(C,90:)`
`cmark_t O_1`	`draw(O1); label(O1,90:)`
`cmark_lb O_2`	`draw(O2); label(O2,-135:)`
`drawcircle O_1 A`	`draw(circle(O1,length(segment(O1,A))))`

Figure 1. Equivalent descriptions of a construction in GCLC (left) and in Eukleides (right) languages.

Figure 2 shows how the description given in Figure 1 would appear in XML version. Notice, again, a direct link between the XML representation and representations in GCLC and Eukleides languages.

Converting from a DGS language to XML would be performed by a specific converter, naturally relying on the DGS's parsing mechanism. Converting from XML to a DGS language will be implemented via an XSLT file.

Having converters from, and to, XML format for all DGSs, we (indirectly) have converters from each format to any other format. Thus, in this way, the base for a common interchange format is provided. XML is a natural framework for such interchange format, because of its strict syntax, verification mechanisms, suitable usage on the Internet, and a large number of available supporting tools.

XML descriptions of constructions can be, by means of XSLT, also transformed into HTML format that is convenient for human-readable display in browsers. It can also be transformed into different representations, such as natural language form.

```
<?xml version="1.0" encoding="UTF-8"?>
<!DOCTYPE figure SYSTEM "GeoCons.dtd">
<?xml-stylesheet href="GeoConsHTML.xsl" type="text/xsl"?>

<figure>
  <draw>
    <dimensions width="80.000000" height="80.000000"></dimensions>
  </draw>
  <define>
    <fixed_point x="10.000000" y="30.000000">A</fixed_point>
    <fixed_point x="60.000000" y="10.000000">B</fixed_point>
    <fixed_point x="50.000000" y="70.000000">C</fixed_point>
  </define>
  <construct>
    <segment_bisector>
      <new_line>a</new_line><point>B</point><point>C</point>
    </segment_bisector>
    <segment_bisector>
      <new_line>b</new_line><point>A</point><point>C</point>
    </segment_bisector>
    <segment_bisector>
      <new_line>c</new_line><point>B</point><point>A</point>
    </segment_bisector>
    <intersection>
      <new_point>O_1</new_point><line>a</line><line>b</line>
    </intersection>
    <intersection>
      <new_point>O_2</new_point><line>a</line><line>c</line>
    </intersection>
  </construct>
  <draw>
    <line>a</line>
    <line>b</line>
    <line>c</line>
    &lt;segment&gt;<point>A</point><point>B</point>&lt;/segment&gt;
    &lt;segment&gt;<point>A</point><point>C</point>&lt;/segment&gt;
    &lt;segment&gt;<point>B</point><point>C</point>&lt;/segment&gt;
  </draw>
</figure>
```

Figure 2. XML version of construction descriptions given in Figure 1.

A specific DTD document would define syntactical restrictions for construction descriptions. This DTD document can then be used, in conjunction with the generic XML validation mechanism, for verifying whether a given description of a geometrical construction is legal.

3.2 SVG Support

As said in the previous subsection, XML format can be used for representing descriptions of geometrical constructions and, hence, as an interchange format for different geometry tools. On the other hand, this representation can be used for visualization of constructions, by using SVG. The visualization data (in SVG) can be generated directly from the XML description of a construction (basically requiring a new geometry tool). Another possibility is to implement, within a DGS, an "export to" SVG option.

With this option implemented, the visualization data (in SVG) could be generated from the XML description indirectly—the XML description would be first converted to a representation of the geometry tool, and then further to SVG format. Note that, with only a limited number of converters, a wide range of processing of geometrical descriptions would be possible.

3.3 Representation of Proofs

Geometrical proofs should be stored in a way that provides:

- strict verification;
- different sorts of presentation for easier understanding.

Geometrical proofs could be stored in different forms, for instance in axiomatic form (e.g., in Hilbert style and sequent calculus style). Representing higher-level proofs, produced by the area method, is also interesting. Proofs generated by this method consist of sequences of equalities involving expressions over geometrical quantities (such as a signed area of triangle). For each step of the proof, leading from one equality to another, there is a detailed justification (in terms of used definition or lemma): for elimination steps, geometrical simplification steps, and for algebraic simplification steps. These proofs have a linear structure, but may involve subproofs (proofs of lemmas).

4 Implementation

In this section, we describe our XML suite for geometrical constructions (see Figure 3) and geometrical proofs. It follows the motivations and ideas given in Section 3 and consists of:

- A newly defined XML-based format for representing geometrical constructions with corresponding DTD; this format covers standard constructions by ruler and compass, but also a range of

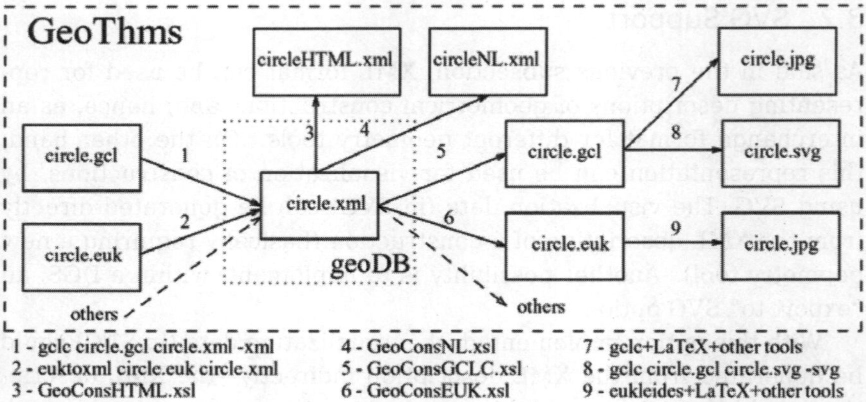

Figure 3. Illustration of architecture of the XML suite for geometrical constructions.

other devices supported by dynamic geometry tools (including compound constructions, transformations, labeling, etc.).

- Converters from dynamic geometry tools to XML-based form (currently, there are converters for GCLC and Eukleides; these converters were written in the programming languages C++ and C, as the main tools themselves).
- Converters for descriptions of constructions from XML-based form to dynamic geometry tools (currently, there are converters for GCLC and Eukleides; these converters were implemented as XSLT files).
- A converter for descriptions of constructions from XML-based form to a simple, readable HTML form (with syntax coloring features, provided for better readability); this converter was implemented as an XSLT file.
- A converter from XML-based form to a natural language form (currently, only for English language); this converter was implemented as an XSLT file.
- A tool for exporting figures from dynamic geometry tools to SVG format (currently, there is a converter for GCLC; this converter was written in the programming language C++, as the main tool itself).
- Newly defined XML-based format for representing proofs of properties of geometrical constructions with a corresponding DTD; the format is adapted for the area method.

- A tool for exporting proofs from automated theorem provers systems to XML-based form (currently, there is a converter for GCLCprover; this converter was written in the programming language C++, as the main tool itself).
- A converter for proofs from XML-based form to a simple, readable HTML form (with syntax coloring features, and other features for better readability); this converter was implemented as an XSLT file.

GeoThms uses an XML format to high extent, for storing, communicating, and presenting data. The presented suite is available via GeoThms (from [url:29]) and partly within a distribution package for GCLC (from [url:133]).

Some examples built with the help of the above-mentioned tools are given in Section 5.

5 Examples

Figure 2 shows the XML code that corresponds to the construction descriptions given in Figure 1. The code is simple and readable. Within the code, there were points A, B, C introduced, and then the bisectors a and b of the sides BC and AC were constructed. The intersection of a and b is denoted by O_1 (note that this point is the center of the circumcircle of the triangle ABC). The four points, the three sides of the triangles, and the circle with the center O_1 containing the point A are shown.

The content of the file shown in Figure 2 is valid with respect to a special-purpose DTD, developed for geometrical constructions. Part of this DTD is shown in Figure 4.

The content of the file shown in Figure 2 was generated by the converter from GCLC to XML format.

The content of the file shown in Figure 2, transformed by the XSLT files `geocons-gclc.xsl` and `geocons-eukleides.xsl`, gives (exactly) the contents in GCLC and Eukleides format shown in Figure 1.

The content of the file shown in Figure 2, transformed by the XSLT file `geoconsHTML.xsl` gives a simple and readable description of the construction presented in HTML. The content of the file shown in Figure 2, transformed by the XSLT file `geoconsNL.xsl` gives a similar description, in HTML, but in a natural-language form (Figure 5).

```
<!--*************constructions*************-->
<!ELEMENT construct (intersection|intersection_cc|intersection_cl|midpoint|
          foot|random_point_on_line|translate|towards|rotate|half_turn|
          line_reflection|inversion|ruler|parallel|perpendicular|
          segment_bisector|angle_bisector|compass)*>
<!ELEMENT new_point (#PCDATA)>
<!ELEMENT intersection (new_point,line,line)>
<!ELEMENT intersection_cc (new_point,new_point,circle,circle)>
<!ELEMENT intersection_cl (new_point,new_point,circle,line)>
<!ELEMENT midpoint (new_point,point,point)>
<!ELEMENT foot (new_point,point,line)>
<!ELEMENT random_point_on_line (new_point,point,point)>
<!ELEMENT translate (new_point,vector,point)>
<!ELEMENT towards (new_point,vector,coefficient)>
<!ELEMENT rotate (new_point,center,angle,point)>
<!ELEMENT half_turn (new_point,center,point)>
<!ELEMENT line_reflection (new_point,line,point)>
<!ELEMENT inversion (new_point,circle,point)>
```

Figure 4. Fragment of the DTD for geometrical constructions.

Description of construction:

Let us draw the following objects:

- **Image dimensions: width 80.000000mm, height 80.000000mm.**

Let us define the following fixed points:

- **Let A be a point with Cartesian coordinates (10.000000, 30.000000).**
- **Let B be a point with Cartesian coordinates (60.000000, 10.000000).**
- **Let C be a point with Cartesian coordinates (50.000000, 70.000000).**

Let us construct the following objects:

- **A line a such that it is a bisector of the segment BB.**
- **A line b such that it is a bisector of the segment AA.**
- **A line c such that it is a bisector of the segment BB.**
- **A point O_1 such that it is a intersection of the lines a and b.**
- **A point O_2 such that it is a intersection of the lines a and c.**

Figure 5. Fragment of a natural-language rendering of the XML file shown in Figure 2.

Definitions:

1 Let M_{a}^{0} be the midpoint of the segment BC.

2 Let T_{a}^{1} be the point on bisector of the segment BC (such that {\tt TRATIO} T_{a}^{1} M_{a}^{0} B 1).

3 Let M_{b}^{2} be the midpoint of the segment AC.

4 Let T_{b}^{3} be the point on bisector of the segment AC (such that {\tt TRATIO} T_{b}^{3} M_{b}^{2} A 1).

5 Let M_{c}^{4} be the midpoint of the segment BA.

6 Let T_{c}^{5} be the point on bisector of the segment BA (such that {\tt TRATIO} T_{c}^{5} M_{c}^{4} B 1).

Step 2

p3(A, O_1, A) = p3(B, O_1, B)

the statement

Semantic values: 2189.795918 = 2189.795918

Step 4

(((segment_ratio(M_{a}^{0}, O_1; M_{a}^{0}, T_{a}^{1})) * (p3(A, T_{a}^{1}, A))) + ((segment_ratio(O_1, T_{a}^{1}; M_{a}^{0}, T_{a}^{1})) * (p3(A, M_{a}^{0}, A)))) + ((-1.000000) * (((segment_ratio(M_{a}^{0}, O_1; M_{a}^{0}, T_{a}^{1})) * (segment_ratio(O_1, T_{a}^{1}; M_{a}^{0}, T_{a}^{1}))) * (p3(M_{a}^{0}, T_{a}^{1}, M_{a}^{0})))) = p3(B, O_1, B)

Lemma 32 (point O_1 eliminated)

Semantic values: 2189.795918 = 2189.795918

Figure 6. A fragment of a proof generated by GCLCprover.

An SVG-based visualization of the construction given in Figure 2 can be obtained by first converting the file shown in Figure 2 to GCLC format, and then by using the option for exporting from GCLC format to SVG.

Our XML suite also has support for storing and presenting geometrical proofs. The current support is aimed only at the proofs produced by the area method (but it is subject to changes and extensions for other proof styles). Consider the construction described in Figure 2. If we construct a bisector c of the side AB, and if we construct the intersection O_2 of the lines a and c, then the points O_1 and O_2 will be identical. This property can be proved by GCLCprover. Figure 6 shows part of this proof presented in HTML. The XML code generated by GCLCprover is simple and readable. It is valid with respect to a DTD

```
<!--******** Definitions **************-->
<!ELEMENT definitions (definition)*>
<!ELEMENT definition (\#PCDATA)>

<!--******** Proof **************-->
<!ELEMENT proof (proof\_step|lemma)*>
<!ELEMENT proof\_step (equality,explanation,semantics)>

<!ELEMENT lemma (proof,status)>
<!ATTLIST lemma level CDATA \#REQUIRED>

<!ELEMENT equality (expression,expression)>
<!ELEMENT inequality (expression,expression)>

<!ELEMENT expression (number|constant|sum|mult|fraction|segment\_ratio|
        signed\_area3|signed\_area4|pythagoras\_difference3|
        pythagoras\_difference4)>
```

Figure 7. A fragment of the DTD for proofs of properties of geometrical constructions.

developed for proofs of properties of geometrical constructions. Part of this DTD is shown in Figure 7.

6 Conclusions and Further Work

We have presented a case for using XML in describing geometrical constructions and proofs, and as an interchange format for dynamic geometry tools. We gave a brief description of the notion of geometrical constructions, XML, and the geometrical software tools that already use our support for XML. Our XML suite is publicly available and used in the GeoThms framework.

This work is related to work in other domains of automated reasoning: joint efforts from a number of researchers have led to standards such as DIMACS (for propositional logic) [84] and SMT (for satisfiability modulo theory) [209]; and repositories of problems such as SAT-lib (for propositional logic) [228], TPTP (for predicate logic) [249], SMT-lib (for satisfiability modulo theory) [209]. Such efforts, standards, and libraries are very useful for easier exchange of problems, proofs, and even program code, and they help advance the underlying field.

We are planning to work on further improvements (based on XML schemes) of the validation mechanism including some semantics checks. Also, we will work on extending and improving the format for proofs, and especially on using applications such as MathML, and on schemes for describing mathematical contents such as OMDoc.

We intend to build further the database of geometrical constructions within GeoThms and, hopefully, to develop it into a major public resource for geometrical constructions, linking a number of geometry tools, formats, and repositories.

From Parametrized Graphics to Interactive Illustrations

M. Kraus

Unfortunately, languages specify graphic positions in an awkward fashion, by numbers.

—Stephen Trimberger [260]

While there are many successful examples of web-based interactive illustrations for publication and education in mathematics and other disciplines, the development of these illustrations still poses many challenges. The diversity of implementations and representations of such illustrations suggests that there is not a single best solution, but a wide variety of potentially successful approaches to the implementation of interactive illustrations. In this chapter, a classification of well-known approaches is presented with a focus on one less popular approach, which is based on parametric descriptions of graphics and the direct manipulation of relevant parameters. The benefits of these parametrized graphics for interactive illustrations are discussed, and the lessons learned from implementing this approach in the Java applet LiveGraphics3D are summarized.

1 Web-Based Interactive Illustrations

Web-based interactive illustrations promise great benefits for teaching and learning in computer-assisted learning environments. Beall et al. [28] define an *interactive illustration* by the requirement of reader participation, i.e., an illustration is considered "interactive if the reader, to obtain the content, must do something more than invoke the mechanisms that present the illustration." Unfortunately, readers[1] often fail to perform the required actions to navigate to, or to initiate the display of, the full content due to the limited affordance of many interactive illustrations. Thus, readers experience frustration and discouragement instead of the aesthetic pleasures potentially provided by interactive works, which are—according to Douglas and Hargadon [87]—immersion and engagement.

Immersion may already be experienced by interactively rotating a three-dimensional illustration in real time (even without stereo viewing). However, apart from the immersive pleasure provided by interactively manipulating the camera parameters, an interactive viewer for rotating three-dimensional graphics cannot significantly enhance the effectiveness of the corresponding noninteractive illustration. In fact, a well-designed rotatable illustration will hardly classify as an interactive illustration since the relevant content is already displayed for the initial view point. It should be emphasized that these considerations only apply to illustrative uses of viewers for three-dimensional graphics in contrast to explorative uses, e.g., for data visualization.

Engagement is different from immersion as emphasized by Douglas and Hargadon [87]: "the pleasure of engagement with hypertext fiction comes from users' access to a wide repertoire of schemas and scripts, our attempts to discover congruencies between the hypertext and an array of often mutually exclusive schemas, and, ultimately, our ability to make sense of the work as a whole." In particular, they observe that readers "who enjoy engagement also tend to enjoy confronting situations for which they lack scripts, as these provide opportunities for learning...." Thus, engaged readers are more likely to draw pleasure from surprises, while immersed readers are more likely to be frustrated by the unexpected.

In the case of interactive illustrations, readers should be (and often are) enabled to interact purposefully with an illustration to test their understanding of the illustrated concept by comparing the effect of their interaction with their expectations. In addition to the (immersive) pleasure of exerting control over the illustration, this provides either the pleasure of an immediate affirmation of the reader's prediction or the surprise of an unexpected reaction, which in some cases already helps to gain a better understanding of the illustrated concept. Thus,

[1]We adopt the notion of a "reader" of an interactive illustration to describe the person looking at an interactive illustration, while the term "viewer" denotes a program to display an illustration.

an opportunity for a stress-free test of the reader's understanding is offered. Complex interactive illustrations can also engage readers in an active exploration of the configuration space of the illustration in order to, for example, examine "special" cases, which are usually avoided when illustrating a more general concept. Remarkably, this kind of engagement can motivate readers to seek surprises by actively trying to disprove their own predictions—and enjoying it!

Unfortunately, many readers fail to experience the pleasures of immersion and engagement because they are unable to interact successfully with an illustration. This is often due to a significant lack of *affordance of interactive illustrations*. Probably the most popular solution to this problem is the use of direct-manipulation interfaces [28]. This approach is particularly successful if specific symbols are consistently used to indicate manipulatable elements of an illustration. As long as readers are aware of them, these symbols may be as simple as points of a specific color, which are often employed in interactive illustrations of geometric constructions; for example, in the Cinderella system by Richter-Gebert and Kortenkamp [214].

An approach closely related to the use of direct-manipulation interfaces in the design of interactive illustrations is the mimicking of traditional illustrations, i.e., the recreation of "the look of a classic textbook" [126]. Apart from helping to overcome mental barriers of some readers who are intimidated by a complex collection of widgets or were frustrated by such interfaces in the past, this approach to the design of interactive illustrations encourages authors to keep illustrations simple, clean, and consistent, or—speaking more generally—to apply design principles that have been established for traditional illustrations. Thus, an interactive illustration can not only convey the content of the corresponding static illustration, but it may actually show the same image initially. In this case, some subtle symbols or a brief textual explanation can indicate that additional content is immediately available to interested readers in the form of meaningful interactions by direct manipulation.

Costs of developing well-designed interactive illustrations are considerably higher than the costs of developing the corresponding static illustrations. Beall et al. [28] describe their experiences as "sobering." Summarizing their comparison, one could estimate that the development of an interactive illustration requires about two orders of magnitude more time than the design of the corresponding static illustrations. Most of the additional time was needed to design, implement, and reimplement the interactive features in compiler programming languages such as C++ and Java. It is important to note that the costs reported by Beall et al. do not include the time necessary in order to learn the programming languages required to implement web-based interactive illustrations. In fact, many illustrators, authors, and educators are lacking these skills—even those who are fluent in other programming languages.

These high costs have motivated many attempts to provide frameworks for web-based interactive illustrations at a more attractive price. The most popular approaches are based on reuse; in particular, by adapting existing Java classes (see, for example, the work by Perlin [203]), employing large libraries of Java classes (e.g., JavaView [204]), or building repositories of interactive illustrations (e.g., Görke et al. [126]). However, none of these approaches has resulted in a breakthrough.

In Section 2, a classification of implementation frameworks for web-based interactive illustrations is presented, which reveals that there are frameworks that allow nonprogrammers to design interactive illustrations at a fraction of the cost of any Java-based programming framework—at least for particular kinds of interactive illustrations, namely, geometric constructions. Obviously, the price of the success of these frameworks is the limitation to a specific application domain. On the other hand, the price to pay for frameworks supporting the implementation of more general web-based illustrations is the requirement to implement the interactive features of an illustration in Java or other languages for web-based programming, e.g., ECMAScript or JavaScript.

An alternative implementation framework for web-based interactive illustrations is based on the concept of parametrized graphics, i.e., the algorithmic specification of graphics in dependency of a set of parameters, and the interactive manipulation of these parameters. Similar concepts were implemented, for example, in the Sam system by Trimberger [260] and in the Fnord system by Banchoff and Draves [88]. In Section 3, the lessons learned from implementing parametrized graphics for web-based interactive illustrations as part of the Java applet LiveGraphics3D [162] are discussed.

The costs of authoring interactive illustrations within this conceptual framework are two-fold: First, a web-based viewer for parametrized graphics has to be implemented; for example, in the form of a browser plug-in or a Java applet such as LiveGraphics3D. Usually, specific viewers have to be implemented for each programming language that is employed to specify parametrized graphics. In the case of Live-

Graphics3D, this is a subset of the programming language of the computer algebra system Mathematica [274]. However, once a viewer for a specific language is implemented, it may be employed to display any interactive illustration specified by parametrized graphics in this language.

The second kind of costs are those of authoring particular interactive illustrations. Ultimately, authors should be able to choose from a set of viewers for various programming languages according to their skills and preferences. The actual costs of authoring an interactive illustration in the form of parametrized graphics will inevitably be higher than authoring a corresponding static illustration in the same programming language. The difference depends strongly on the author's programming skills and the degree of sophistication of the desired interactive features of the illustration. For experienced programmers, who are used to designing static illustrations in a particular programming language, the overhead might result in an authoring time about twice as long as for a noninteractive illustration[2].

2 Classification of Frameworks

There is a wide variety of published frameworks for the implementation of web-based interactive illustrations. Since these systems are published in several independent research communities, it is quite difficult to gain an overview of the relevant publications. Thus, the classification presented here is without doubt incomplete; in fact, no attempt is made to include all relevant systems and frameworks but only one or two representative examples per category, with a preference for frameworks supporting three-dimensional illustrations since two-dimensional illustrations may usually be implemented within these systems, too. Moreover, the interaction with three-dimensional illustrations offers particular challenges, which are difficult to handle in frameworks focusing on two-dimensional graphics. Frameworks based on de facto web standards (such as X3D, VRML, ECMAScript, JavaScript, and Java) were preferred since systems based on proprietary formats and web browser plug-ins are often limited to particular operating systems and/or web browsers. While this is acceptable in certain learning and teaching scenarios, many web-based illustrations are significantly less attrac-

[2]This estimate is based on my personal experience with adapting more than 60 interactive illustrations [165] from a textbook on CAGD [114]; on average, each took about one to two hours.

frameworks for interactive illustrations				
event-driven graphics		constraint-driven graphics		parameter-driven graphics
interactive graphics programs	graphics enhanced with virtual sensors	geometric constructions	graphics with one-way constraints	

Figure 1. The proposed classification hierarchy.

tive because readers are required to download and install (or even to buy) a browser plug-in before accessing the illustration.

The proposed classification scheme is illustrated in Figure 1. On its top level it distinguishes between event-driven, constraint-driven, and parameter-driven graphics. For event-driven graphics, event handlers are specified by the author or—more precisely—the programmer of the graphics. For constraint-driven graphics, constraints between variables, coordinates, and attributes of graphics elements have to be specified. Thus, authors of constraint-driven graphics usually do not need to explicitly program event handlers. However, the actual implementation of constraint-driven graphics may employ event handlers internally. For parameter-driven graphics, the algorithmic dependencies of coordinates and attributes on user-defined parameters have to be specified. The first two of these top-level categories are divided into subcategories specific to web-based interactive illustrations. More details about each subcategory and examples of frameworks supporting each kind of approach are provided below.

Event-driven graphics define events and corresponding event handlers, which are executed if the specified event occurs. Events may be user actions on any abstraction level, e.g., mouse button clicks or the entry into a particular room of a virtual environment. Two subcategories are distinguished: general interactive programs and graphics formats with virtual sensors.

Interactive graphics programs are implemented in general-purpose programming languages and render a graphical output depending on the user's input. For web-based illustrations, the most important example for this way of specifying interactions is that of Java applets. The required Java classes are sometimes implemented from scratch, but more often they start from similar classes (in Java or other programming languages); see, for example, the discussion by Beall et al. [28] or Perlin [203]. There are also libraries of Java classes for many common tasks related to the implementation of three-dimensional interactive illustrations, e.g., JavaView [204] or Java3D. Moreover, projects have been started to build repositories of Java applets for interactive illustrations, e.g., by Görke et al. [126].

Java applets—and interactive graphics programs in general—provide the greatest flexibility but result in significant development costs. Moreover, the development environments of general-purpose compiler languages are usually less appropriate for the iterative design of interactive illustrations, as noted by Beall et al. [28]. Apart from the high development costs, the programming skills required to implement interactive illustrations this way are a major hindrance to potential users.

Graphics enhanced with virtual sensors include the specification of user interactions in scene graphs with the help of particular nodes, often called "sensors." These sensors react to certain events by executing an associated event handler. The web's standard format for scene graphs of three-dimensional graphics is VRML and its successor X3D. These graphics formats allow for sensor primitives that may call event handlers, which are defined, for example, in ECMAScript. Similarly, scripted event handlers can be associated with two-dimensional graphics defined in SVG.

This sensor mechanism is most useful if predefined event handlers may be employed and if only a small part of the whole scene graph is affected by interactions. However, in many cases there are no adequate predefined event handlers for a particular interactive illustration. Moreover, it is often less costly for authors to implement the construction of the whole illustration from scratch after each interaction rather than modify only certain parts of it; in particular, if almost all parts of the illustration are (potentially) affected by one or another interaction and the rendering performance is no issue.

While static three-dimensional graphics may be designed with interactive graphics tools and converted to X3D or VRML, this approach turns out to be rather inefficient for many interactive illustrations since there are often only very few static elements in interactive illustrations, especially in illustrations of abstract concepts. On the other hand, the dependencies on many user interactions can only be integrated with the help of scripted event handlers without the advantages of graphical editing tools. As scripting languages are usually less expressive and their performance limited, it is often preferable to implement interactive illustrations with the help of a general-purpose programming language.

Constraint-driven graphics are suitable for interactive illustrations that may be implemented by satisfying certain (in most cases geometric) constraints while the reader interactively manipulates parts of the graphics. The most important advantage for authors is the possibility to define and edit many constraints with the help of a graphical editing tool. In this case, no programming whatsoever is required to design an interactive illustration. The first implementation of such a system was Sketchpad by Sutherland [250]. Constraint-driven graphics may be further classified by the kind of supported constraints.

Geometric constructions are easily specified with the help of geometric constraints in interactive graphics tools without any programming; thus, constraint-driven graphics are extremely popular for the design of interactive illustrations of geometric constructions and proofs. There are several published systems and commercial products; two examples based on Java are JavaSketchpad [157] and Cinderella [214].

The most important limitation of this approach is its restriction to geometric constraints. The theoretical foundation of Cinderella indicates that attempts to adapt these frameworks for more general constraints is extremely challenging. Additionally, the editing of more general constraints in a graphical tool is problematic. Therefore, the remarkable success of constraint-driven graphics might actually be limited to geometric constraints.

Graphics with one-way constraints avoid the difficulties of satisfying general constraints by employing only a very particular kind of constraint: instead of computing a global solution satisfying all constraints, each constraint is satisfied individually by adapting a single variable associated with the constraint. In some cases, it is possible to create a tree of dependencies such that this approach results in a global solution; however, in general, this approach will not result in the satisfaction of all constraints; thus, authors and readers are often surprised by the resulting graphics. This complicates the design of interactive illustrations considerably; Vander Zanden et al. [262] discuss many of these problems in detail.

Moreover, specifying general constraints often requires a level of expressive power that is only found in programming languages. Usually, these constraints cannot be satisfied automatically; thus, authors also (or exclusively) have to supply program code to satisfy the one-way constraints. Obviously, there is a smooth transition from satisfying a one-way constraint to the execution of an event handler supplied by the author. Examples of published web-based frameworks for one-way constraints are VRMLC [82] and CSVG [18].

Parameter-driven graphics (also called "parametrized graphics") are based on the algorithmic definition of graphics depending on user-specified parameters. In this approach, authors of interactive illustrations specify the initial values, which may later be modified by readers interacting with the graphics. Any such change of a parameter results in the re-evaluation of the graphics. Thus, parameter-driven graphics may sometimes resemble a single event handler that builds the graphics from scratch. Alternatively, parameter-driven graphics may also be considered to specify one-way constraints for all coordinates and attributes that depend on user-defined parameters. Note that the graphics formats for parameter-driven graphics have to be considerably more expressive than standard graphics formats. Moreover, the display requires frequent reevaluations of the graphics in addition to the rendering.

Early work on parameter-driven graphics includes Trimberger's system Sam for the simultaneous editing of a graphical, and a textual representation, of the same graphics [260]. The intention was to reflect any modification of the textual representation instantaneously in the graphics display and vice versa. The project was very ambitious, in particular considering the computational resources available at that time, and raised several important questions.

Another system—the Fnord system—which is also related to parameter-driven graphics, was developed by Draves [88] and employed by Thomas Banchoff for mathematical visualizations. Beall et al. [28] characterize it this way:

> Fnord is an expressive language, but interaction with it is through a pre-defined metaphor: one of 'adjustable constants.' Certain entities within the language can be defined as the value of a function called *Slider*, so that one might write *b* = *Slider(0, 100, 50, 'curvature')* to indicate that there should be a slider placed on the screen.... When the user adjusts this slider, the value of *b* is changed, and all things that depend on *b* are updated as well. This, together with a kind of virtual sphere interaction for changing viewpoints within 3D windows, constitutes the entire interaction model. Because of this, Fnord can only provide a limited set of possible interactive illustrations.

According to the Fnord manual [88], this description is not quite correct since there is also a "point widget" available for user input of two-dimensional and three-dimensional points. Thus, direct manipulation of parameters is, in fact, available. This is certainly still not adequate to allow for all possible interactive illustrations. However, entering numbers by dragging points is an extremely powerful interaction technique. Unfortunately, Fnord uses its own language—in some regards similar to a computer algebra system but not quite as powerful—to specify graphics, which presumably has strongly limited its popularity.

Independently of this earlier work, a parser and evaluator of a subset of the programming language of the computer algebra system Mathematica [274] was implemented in Java by a group of undergraduate students in 2000. The students' code had to be reimplemented for performance reasons and was integrated into LiveGraphics3D [162], which is a previously implemented Java-based viewer for three-dimensional Mathematica graphics. Additionally, several extensions of the initial subset of the language were necessary. Moreover, the direct manipulation of parameters had to be designed and implemented. Nonetheless, a first version of this implementation was presented in 2001 [163,164]. Since then, it has been employed in several published projects by Mathematica users, e.g., Rogness [219] and Mora [190].

3 Interactive Illustrations in LiveGraphics3D

There is, in fact, a continuously growing interest in LiveGraphics3D and the use of parametrized graphics for web-based interactive illustrations—in particular, within the Mathematica users community. Some of the lessons learned from this project since 2001 are summarized in this section.

Beall et al. [28] emphasize that the development environments of many compiler languages are not well suited for the rapid prototyping often required for the design of interactive illustrations. Therefore, they recommend developing a first prototype in an interpreted system. This observation motivated the idea to exploit the expressive power of the programming language of a computer algebra system to specify parametrized graphics. In contrast to previous systems such as Sam and Fnord, a carefully chosen subset of the programming language of a popular computer algebra system, namely, Mathematica, was chosen for the implementation of parametrized graphics in LiveGraphics3D. This ensures that Mathematica can be employed to prototype illustrations efficiently, i.e., an interpreted development environment is already available without any additional development efforts. Moreover, many Mathematica users are already familiar with this programming language as well as with the development environment.

It should be emphasized that once developed with Mathematica, the interactive illustrations are displayed and evaluated by LiveGraphics3D independently of Mathematica; i.e., the applet is supplied with a textual representation of the parametrized graphics, which is parsed, evaluated, and displayed by the applet without any communication with Mathematica. Of course, the textual representation of the parametrized graphics may also be generated without Mathematica. In the latter case, Mathematica is never needed at any point in the process.

While the decision to employ the programming language of a popular computer algebra system to specify parametrized graphics in LiveGraphics3D turned out to be extremely beneficial; there are, however, also disadvantages of this approach. One disadvantage in particular is the restriction to a specific programming language and, therefore, to users who are already skilled in this language. Additionally, the chosen language was neither designed for parametrized graphics, nor for interactive illustrations, while a new language design certainly could have been considerably more elegant. Similarly, there are also difficulties when using Mathematica to design and prototype interactive illustrations.

Overall, however, the feedback of Mathematica users to the concept of implementing interactive illustrations with the help of parametrized graphics was quite positive. In particular, several users expressed their surprise at the real-time performance of the evaluation of basic Mathematica expressions in a Java applet, and there are already first publications about specific applications [190, 219].[3]

With respect to the technical aspects of this approach, the most important lessons to be learned were these:

- While Mathematica's language features a large amount of functions and structures, it was possible to find a rather small subset for the implementation of parametrized graphics in LiveGraphics3D.
- Originally, LiveGraphics3D was designed to display unparametrized graphics; therefore, it was decided to limit the parametrization to coordinates of points since the parametrization of attributes of graphics primitives would have required significantly higher development costs.
- For performance reasons, LiveGraphics3D never recomputes the lighting of a scene. For static graphics this is fine, but for parametrized graphics this is definitely a shortcoming.

[3]There are considerably more publications about projects using LiveGraphics3D as a viewer for unparametrized graphics.

- For interactive illustrations in three dimensions, two-dimensional graphics overlaying the rotatable three-dimensional graphics would be very useful to emulate widgets such as sliders. However, this feature had never been implemented in LiveGraphics3D since it is considerably less important for unparametrized graphics and would require rather high development costs.
- If interactive illustrations are used for teaching and learning, annotations become crucial and, therefore, the typesetting of mathematical symbols in LiveGraphics3D had to be improved substantially.

Thus, applying and enhancing LiveGraphics3D to evaluate, render, and directly manipulate parametrized graphics for interactive illustrations revealed several shortcomings and limitations of the applet, which are technically unrelated to the direct manipulation of parametrized graphics and were considered acceptable for static graphics but turned out to be significant shortcomings in the context of interactive illustrations. In fact, it is quite likely that viewers for "general" three-dimensional graphics are almost always designed and implemented with certain application scenarios in mind. Therefore, successfully extending a graphics viewer for interactive illustrations will often include enhancements that are technically unrelated to the interaction with graphics but are just as important to display well-designed illustrations.

4 Suggestions for Future Developments

Web-based interactive illustrations have been implemented successfully within many frameworks, which may be classified as event-driven, constraint-driven, and parameter-driven. There is surprisingly little published work on interactive illustrations based on the latter approach, i.e., parametrized graphics, probably because of the lack of direct manipulation of parameters in most viewers for parametrized graphics. However, the implementation of parametrized graphics in LiveGraphics3D proves that a subset of the programming language of a popular computer algebra system may be used without modifications to specify parametrized graphics. Moreover, parameters in these graphics may be directly manipulated within this web-based viewer in real time.

Based on these experiences, it is recommendable to implement similar web-based viewers for programming languages of popular computer algebra systems, e.g., Maple or MuPAD. Many graphics languages, e.g., X3D, VRML, and SVG, lack the expressive power necessary for parametrized graphics unless they are enhanced with event handlers implemented in programming languages such as ECMAScript. An important exception is the language of MetaPost [136], which is expressive enough to solve systems of linear equations and is used by several authors to design illustrations. One of them is Donald Knuth, who stated in a panel discussion [258]:

> I'm a big fan of MetaPost for technical illustrations. I don't know anything that's near as good, so I'm doing all the illustrations of *The Art of Computer Programming* in MetaPost. Also, the technical papers I've written are going to be published in a series of eight volumes by Cambridge University Press, and all the illustrations, except the photographs, are going to be MetaPosted.

Implementing frameworks based on parameter-driven graphics for the language of MetaPost and other programming languages used for illustrations would result in very convenient tools for interacting with illustrations defined in these languages. In summary, the most important lesson to be learned for the design of such frameworks is a very simple one: *Know your user!* Not everyone is interested in authoring interactive illustrations in a formal programming language; and of those, who are used to designing technical illustrations in this way, only a small minority is willing to switch to (and learn) another programming language for this task. However, many authors and educators would highly appreciate tools that could help them to enhance their illustrations with meaningful interactions if the costs are reasonable.

3 Educational and Cultural Frameworks

Reaching Mathematical Audiences across All Levels

T. Banchoff

In the Digital Era, all aspects of communicating mathematics are undergoing very rapid changes and it is difficult to make predictions about the future on the basis of present observations. Nonetheless, some things are clear: interactive Internet-based communication of mathematical ideas offers great possibilities for engaging larger and larger audiences in many different ways, for the general public, and for students at all levels. In this chapter, we present some examples of ways to expand communication to these two audiences and we speculate about future developments. The first part of the chapter treats communicating mathematics to the general public, and the second part is a report on an ongoing project developing courseware for Internet-based teaching of undergraduate mathematics courses.

1 Communicating Mathematics to the General Public

For the general public, we can introduce electronic metaphors for traditional ways of presenting ideas, for example, the art gallery exhibit, the traveling exhibit, the public relations poster as Advent calendar, enhancements of published volumes, and anniversary sites.

1.1 Gallery Exhibitions, Actual and Virtual

A mathematical art exhibition traditionally has been a real event, in a particular physical space over a specific time, designed to introduce a

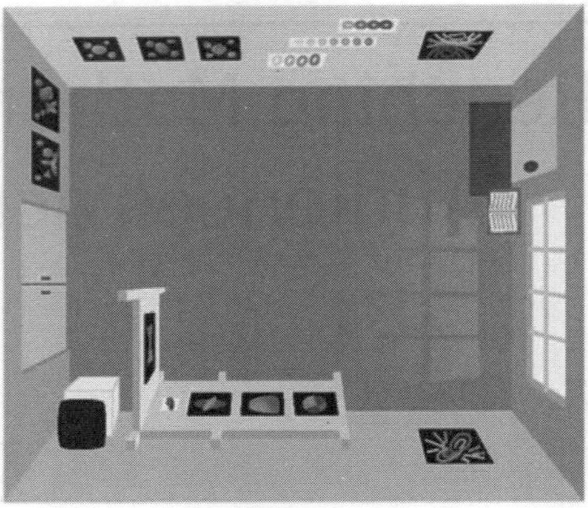

Figure 1. Surfaces beyond the third dimension: gallery view.

broader audience to engaging mathematical imagery. Now, in the Digital Era, such exhibitions can take on a new form, as virtual recreations of actual exhibits, or even as totally virtual constructions. One example of such a recreation of an actual exhibit is "Surfaces Beyond the Third Dimension," first presented by the author and Davide Cervone at the Dodge House Gallery of Providence Art Club in Providence, Rhode Island, in 1996. The process leading up to the virtual version is described in the article [22] and the virtual exhibit itself is accessible at [url:135].

1.2 The Traveling Exhibit "Para Além da Terceira Dimensão"

The electronic gallery book of "Surfaces Beyond the Third Dimension" gave visitors from many countries a chance to sign in and make comments. One comment in particular, in December 1999 from José Francisco Rodrigues at the University of Lisbon, led to a different kind of collaboration resulting in the creation of a traveling exhibit, described in detail in the article [23] and accessible online at [url:2]. It consisted of a number of large panels that could easily be shipped to various locations and displayed in different exhibit configurations, often along with videos and interactive terminals. For two years starting in 2000, the exhibit appeared in a dozen different institutions in Portugal, as well as at the twenty-third Colóquio Brasileiro de Matemática, held in IMPA, Rio de Janeiro, July 2001. A full report of the itinerant exhibit is found at [url:3].

1.3 The "Math Awareness Month" Poster as an Electronic "Advent Calendar"

In 2000, the author served as president of the Mathematical Association of America and as the chair of the Joint Policy Board for Mathematics (the MAA together with the American Mathematical Society and the Society for Industrial and Applied Mathematics). Each year the JPBM selects a theme for a poster and ancillary materials to be distributed to all college and university mathematics departments, concentrating on one particular topic in mathematics that is intended to appeal to a broad student audience.

The theme for April 2000, Math Awareness Month, was "Math Spans All Dimensions." The poster featured a central image presenting the dimensional analogy, from zero-dimensional points to one-dimensional curves, then surfaces with two dimensions and solid spaces with three. At various parts of the poster, photographs of particular individuals appeared, along with some words indicating their connection with the theme. The paper version of the poster was distributed to mathematics departments at all colleges and universities in the United States, as well as to more than 100,000 members of the National Council of Teachers of Mathematics.

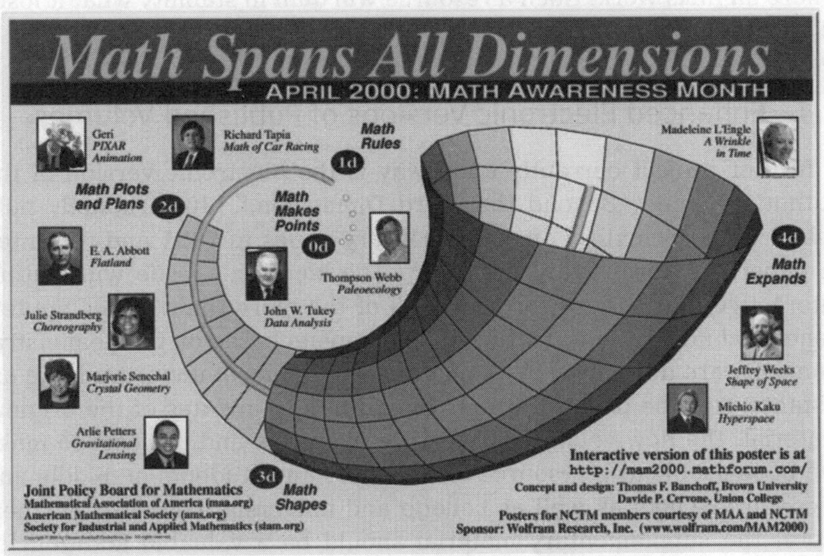

Figure 2. MAM2000 web poster with hyperlinks.

The real power of the poster was the invitation to visit the electronic version for an enhanced experience, accessible through [url:134]. Selecting a photograph on the computer screen opened up a rich collection of materials that expanded the particular contribution of each of the featured contributors, including links to other sites. (The metaphor for selecting an image and finding something behind it that leads to further exploration is an Advent Calendar, with one image for each day leading up to Christmas. Opening the window for a particular day will show an image and some text, in preparation for Christmas Eve.) In the case of the dimensional poster, a viewer could not only investigate the particular topic for each of the individuals featured on the poster, but he or she could also go more deeply into the ways different dimensions appear in a variety of contexts, and into the ways that mathematicians and their collaborators work together to investigate and portray mathematical relationships.

The dynamic nature of the public relations project for mathematics made it appropriate to link to remote sites rather than providing access to a permanent site maintained by the MAA or some other archiving agency. Unfortunately, linking to contemporaneous sites has led to a predictable and inevitable deterioration of the sites featured in the 2000 poster, the phenomenon of "link rot." A possible response to this situation is a self-contained version of the electronic poster on a CD, where all links work. Such a resource will gain in stability what it loses in spontaneity.

1.4 Enhanced Electronic Versions of Published Volumes

A further project currently underway is the electronic version of the author's volume "Beyond the Third Dimension," [20] originally published in the Scientific American Library series in 1991 and reprinted in paperback form five years later. The electronic issue will feature interactive illustrations for nearly all of the more than 100 computer-generated illustrations, featuring Java applets based on the demonstration software developed by the author and teams of undergraduate assistants over the past ten years. In addition to updates of the original material, the new volume will feature exercises and projects to make the book usable as a resource for teachers and students in middle and secondary school as well as college and university courses. Furthermore, the supplementary material should be a valuable addition for the original audience of scientifically interested laypersons. Many students have been involved in the creation of these images and interactive demonstrations, in particular, Davide Cervone, David Eigen, Dan Margalit, Nicholas Thompson, Linda Gruendken, and Michael Schwarz.

1.5 The Electronic Anniversary Book

A final example is the electronic update of the idea of an Anniversary Book associated with a birthday celebration, in this case the author's 65th. Former students Davide Cervone and Ockle Johnson constructed a website that features entries to all of the sites mentioned above, as well as a collection of pictures and personal histories of several dozen undergraduate student assistants over nearly 40 years of teaching. This site can be accessed at [url:136].

These five examples present evidence to support the claim that interactive computer graphics is changing the way mathematicians can relate to the general public.

2 Internet-Based Teaching and Learning in Mathematics

This section provides an update of the report given at the Symposium on Technology in Education at KAIST in Korea in May 2005 [21], which has been reissued by the American Mathematical Society. The focus of this report is an exploration of ways in which the Internet can facilitate communication between teachers and students, and among students. Such courseware developments can make more pedagogical sense out of different time slots, in particular, for classes that meet twice each week. Other issues discussed in this section are student privacy, and forms of assessment. We present here some evidence obtained from the latest version of our online courseware, which was used in a second-semester calculus course for beginning engineering students in the fall semester of 2006.

2.1 Communication Software for Assignments and Examinations

Regarding student privacy, there was the issue of private versus public postings. In our courseware, students are expected to hand in their homework and take-home tests online. Each of the assignments and tests has a time-release feature so that up to a point, the work submitted by the student is available only to the student and to the instructors, and after that point, everyone in the class can see what everyone else has done. Until this past semester, the student had no control over access to the submitted responses to homework or to tests. In our evaluations of the previous versions of the course, we asked about satisfaction with this arrangement, and most respondents indicated that they had no problem with it. A small number said that they were not comfortable either with other students reading their homeworks or their tests or both.

As a result of student feedback, we instituted a new feature, to make it possible for any posting to be designated as "private," to be viewed only by the person posting it and any instructor. A response to a private posting is automatically designated private. It is possible to modify the designation at a later time. When we instituted this modification, we expected that students would be more likely to use the privacy feature in the beginning of the course, until they became familiar with the procedures and until they became comfortable with the idea of having others view their work, even when it is not in completely polished form. Preliminary analysis of questionnaire responses for the fall 2006 second-semester calculus course indicated that this is, indeed, the case. Students were all comfortable with other students having the ability to look at their submitted homework, especially because the privacy option was always possible.

We anticipated that some students would feel differently about having other students see their work on examinations, but most seemed to feel the same way about examinations as they did about homework, considering the fact that the grades themselves were always private.

In previous semesters, a number of students expressed dissatisfaction with the idea that they would be required to type in all of their homework using HTML, sometimes with macros that simplified the typing of mathematical expressions. Students often reported that they would work out the solutions to problems on paper and then type out the solutions into the homework tensor, a laborious process. Although they appreciated the immediate feedback from the instructor, many students did not see much value in looking at the work of other students directly, although some specifically appreciated the selected student work featured in the Solution Key.

In response to this reaction, we continued to try to make it easier to enter mathematical expressions, and at the same time we tried to facilitate the process by which students can upload their paper solutions. Preliminary indications are that this has worked very well. Some students reported that they would type in certain kinds of answers and scan others. Some started typing everything and then switched to working on paper and scanning their submissions. One or two commented explicitly that the bother of typing or scanning was compensated for by the chance to get an immediate response to their work.

In many institutions, it is mandated that instructors distribute evaluation forms for students to complete at the end of each course. Most often the questions asked are based on a standard model, where teachers assign homework that is collected and corrected and handed back at some later time. Availability of the instructor is determined by the frequency and convenience of scheduled office hours. The online model changes this in that the instructor is always available for questions about homework problems, and students frequently receive feedback during the process of solving problems and entering their solutions. Instead of asking questions in class, for example, about readings, students can express their concerns online. Just as a question in class can elicit a response, the same is true online, except that the instructor has more time to consider the question and come up with a response that is useful for the class as a whole. A possible way of handling this is to have a standard "questions" slot after every lecture, with room for a summary of what occurred in class and places for supplementary material.

2.2 Visualization Software in the Classroom and on Assignments

The other important value of online homework is the chance to utilize visualization software and to share insights and results with the rest of the class. It is possible to consider families of mathematical objects like curves and surfaces, to see them as control-response situations. Making observations and formulating conjectures is part of mathematical thinking.

Two illustrations will show how this takes place in a class. In this course, the second class meeting was devoted to a discussion of the courseware and the labware. Although everyone was able to complete the questionnaires and the autobiography, not everyone tried the lab questions. In the future, we will require everyone to investigate lab examples and save results, placing demonstrations and pictures and drawings inside homework entries.

Not many students participated in this way. The ones who did were not always able to coordinate their geometric insights with their algebraic calculations. A particular example is the work of RH, who im-

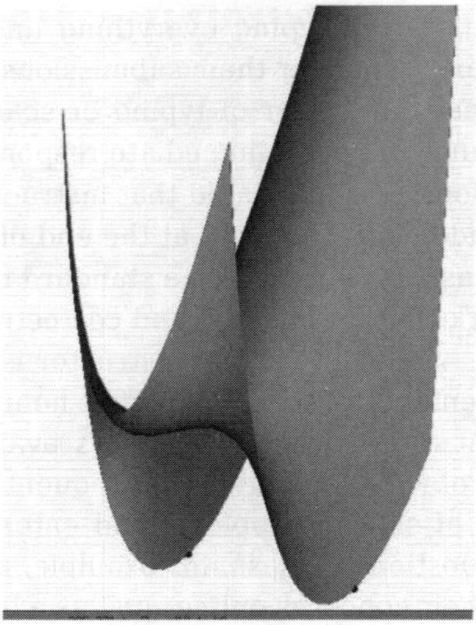

Figure 3. Graph of a function over a rectangle with minima incorrectly indicated.

mediately began using images in her work. She learned right away to mark points on a graph that were the candidates for local maxima and minima, and to color differently the absolute extrema. In one key problem, however, she neglected to see that there was a discrepancy between the two representations. The points that she indicated on her diagram were evidently not the actual minima. That should have led her to check the equations in the demo, and then to check the calculations. In any case, there should have been an observation that there was a difficulty. (See Figure 3 for the plot of an example function under consideration.)

Students did not see the relevance of the graphing software to the solution of what was perceived as an algebraic problem, namely, finding the zeros of the function $f(x,y) = yx^3 - xy^3 = xy(x^2 - y^2) = xy(x+y)(x-y)$. Part of the problem is logical—the product of four expressions is zero if and only if one of the expressions is zero. Graphing the function together with the horizontal plane indicates that the solution is four lines through the origin (see Figure 4).

Finding critical points of a family of functions is aided by visualizations. The complete solution should show pictures of the "before and after" situations for any singular value, where the control space is an interval. For a two-parameter family, the situation requires a further analysis. An example of the first is: $f(x) = x^3 + ux$, and an example of

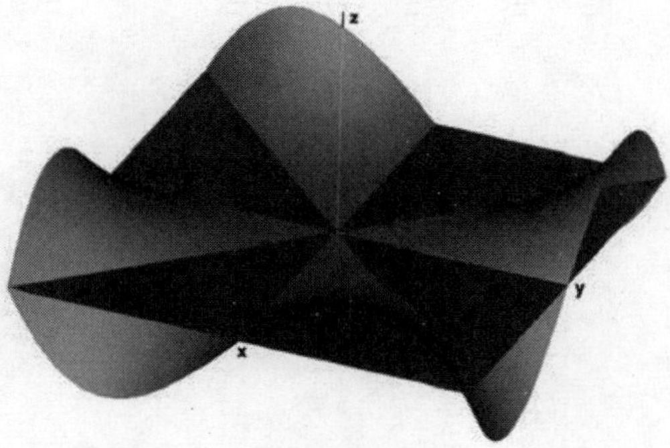

Figure 4. Graph of $z = yx^3 - xy^3$ showing regions where the function is positive.

the second is: $f(x) = x^4 + ux^2 + vx$. The basic idea here is that students should carry out investigations in the control space, make careful observations, and then formulate conjectures. The observation phase can be helped by providing worksheets or tables, and the visual record can be facilitated by allowing students to mark a region by selecting points in a line or a plane. For points on a line, the addition of a point could cause a color change on the complementary intervals. In the plane, it is the introduction of a closed curve that causes the same kind of color switching. This can be effective in topology as well as in calculus, as many demonstrations are.

We continue to make modifications to our courseware based on responses of students who have taken courses, and on comments of colleagues who have heard presentations on this experimental project on Internet-based Teaching and Learning.

3 Summary and Conclusion

In the Digital Era, communicating mathematics will continue to challenge our efforts on all levels, in all aspects of our profession. Conferences like CMDE2006 are important places to share experiences and to point the way to the future.

Toward Autonomous Learners of Mathematics

Olga Caprotti, Mika Seppälä, and Sebastian Xambó

In the context provided by the title of this book, the aim of this chapter is to reflect on the importance of fostering autonomous learning of mathematics by means of available technologies, to reflect on the main issues that are relevant for that purpose now and in the coming years, and to discuss the bearing on such questions of some of the developments produced by the WebALT project.[1]

1 Background

They know enough who know how to learn.

—Henry Brooks Adams (1838–1918)

The mission of teaching, of mathematics in our case, is to catalyze adequate learning by students.

By "students," we mean all people that are required to take courses with mathematical content, from high school to colleges and universities. In the case of colleges and universities, the great majority of these students are studying for science and engineering degrees. For example, of the 30,000 students at the Universitat Politècnica de Catalunya (Technical University of Catalonia), only about two percent are mathematics or statistics majors, but all of them are required to take one or more semesters of mathematical subjects.

[1] European e-Content project "Web Advanced Learning Technologies," Contract Number EDC-22253.

By "adequate learning," we mean that the students must acquire, along with knowledge of the relevant subject matter, a number of competencies related to the corresponding degree. In higher education, for instance, the European Commission specifies that the Diploma Supplement should contain "a precise description of the academic career and the competencies acquired during the study period," and "an objective description of the [student's] achievements and competencies." Among the competencies, *critical thinking* is usually regarded as the most fundamental. Its development ties in so well with mathematics and its applications that mastering these goes a long way toward an effective realization of the paramount "learning to learn" capacity.

The teaching of mathematical subjects, therefore, should aim at eliciting from the students the habitual practice of good critical thinking in the context of the current subject matter. The question here, then, is in what ways can the "digital era" assist in these endeavors. Of course, the implicit message of the book title is that it should make a real difference, at least in "communicating mathematics."

1.1 Are the Old Ways Still the Best?

Some think they are. According to [156], the recipe to fix public education is perhaps simpler than what it could possibly be: "A teacher, a chalkboard and a roomful of willing students." In more detail:

> The plain truth is we need to return to the method that's most effective: a teacher in front of a chalkboard and a roomful of willing students. The old way is the best way. We have it from no less a figure than Euclid himself. When Ptolemy I, the king of Egypt, said he wanted to learn geometry, Euclid explained that he would have to study long hours and memorize the contents of a fat math book. The pharaoh complained that that would be unseemly and demanded a shortcut. Euclid replied, "There is no royal road to geometry."

Admittedly, there are strong reasons to defend the role of good Socratic teachers (assuming those in front of the chalkboard are such). They may indeed be very effective in bringing their students to embrace critical thinking. It has to be remarked, however, that in the present-day circumstances, their impact on large crowds of students must be very limited unless they have means for amplifying dramatically their capacity for interaction with them. It seems safe to say that even the most convincing Socratic teacher cannot do a proper job today (as for example in engineering and science schools) if she or he cannot rely on an environment capable of interacting *autonomously* with most of the students for a good part of the time, in ways that affect them as the direct teacher's contact would.

Since such environments cannot be implemented without advanced technology, the conclusion is that for the job of properly teaching the large number of students taking mathematics courses, there is an ever-increasing need for solutions that boost and magnify the teachers' capacity to inspire and coach their students. Fortunately, we believe that such solutions are slowly becoming available, and we can imagine teachers finally endowed with the capacity to provide a sort of "royal road" to the learning of their subjects.

1.2 Origin of the Digital Era: The Three Main Insights of Shannon

In analyzing what the digital era can bring to the teaching and learning of mathematics, it may be useful to outline the key theoretical advances that made it possible. These advances are basically due to Claude Shannon, and in retrospect they appear with a miraculous aura, for they were generally deemed unfeasible before he formulated his mathematical theory of information (see, for example, [235], which contains a reprint of the original paper [234]). Incidentally, it is interesting to ponder, in connection with "Communicating Mathematics in the Digital Era," that the title of Shannon's far-reaching landmark work is "The Mathematical Theory of Communication" (it was only later that other authors replaced "communication" with "information").

It is remarkable that Shannon's theoretical breakthroughs, brought forth by his deep mathematical insight, were conceived and established long before we had word processors, digital music, and cellular phones; digital cameras and digital TV; or PDAs combining a variety of powerful information and communication features. His theories were clearly stated and proved before anyone could dream about the multimedia world and its great potential in teaching and learning (cf. [39]), and are briefly summarized in the next few paragraphs.

Principle of digitization. Text, sound, and images (including video) can be represented by a string of bits in such a way that each can be reconstructed from its representation and is indistinguishable from the original by the human senses.

The working of this principle is straightforward for text, since text is a string of characters and characters can be encoded (say in the Unicode standard) for practically all writing systems (see [url:176]). Notice, however, that the final visual rendering of the text characters amounts to manipulations of digital images.

The principle is a bit more involved for sound. In this case, it was shown by Shannon that it is enough to sample the air pressure at twice the maximum frequency that is audible by the human ear (about 20 KHz), and to quantize each sample by (say) a 16-bit binary number (this amounts to 65,536 possible intensity levels, and in the near future, the standard number of levels will be more than 100 times higher).

For images, the sampling units are the *pixels*, and the quantizing is done for each of the three fundamental colors. Video is represented as a sequence of digital images. Again, if the pixels are sufficiently small, as in high definition TV or high resolution cameras, then the human eye cannot discriminate the original image from the image reconstructed from the binary representation.

Optimal compressibility. The raw stream of bits delivered by the digitization process usually contains much redundancy, as for example, in areas of uniform color in an image. Shannon found a precise way to measure this redundancy, or (equivalently) the information content of the stream, and proved that it can be compressed into another binary sequence that has the same information as the original one and that cannot be further compressed without loss.

Possibility of fast transmission without errors. When binary-coded information is sent from one place to another, it is transported by some system that is generically called the *channel*. This channel has a *transmission rate*, say, in bits per second. But the "noise" in the channel corrupts some proportion of the bits sent, in a random way, and thus there is an information loss. To protect against this loss, there is the notion of error-correcting codes. The idea is to add redundancy to the information before transmission in such a way that it can be used at the receiving end of the channel to discover the corrupted bits. The problem is that in this way, the information transmission rate is lower, and for a long time it was believed that smaller and smaller proportions of bit errors could only be achieved by lower and lower information rates. Shannon's celebrated channel-coding theorem, a really fundamental breakthrough, states that it is possible to use coding schemes that guarantee as small a proportion of erroneous bits as wanted and yet have an information rate as close to the channel rate as desired.

1.3 The Other Two Engines of the Digital Era

The realization of Shannon's theoretical insights has been made possible by parallel developments that are at the heart of the digital era and that have to do with computing and communication networks.

Increasing computing power. We have witnessed the spread of ever more powerful computing devices at lower and lower prices. As recently noted by a leading expert in the field of computer architectures, there is more computing power in one modern cell phone than in any computer at the time Armstrong went to the Moon. It is foreseen that the evolution in the near future will follow similar trends, to a great extent driven by advances in parallel computing.

Wideband networks. Increasing bandwidths (which amount to higher transmission rates) and decreasing access costs, have been, and will continue to be, the dominating trends in communications networks. The blending and synergy of these technologies with computing, which among many other things make possible remote and distributed computing, are to a great extent the traits that distinguish the digital era from any era in the past.

In this world, the three main insights of Shannon work at their fullest. Terabytes of digital information are daily compressed and decompressed, sent through high-speed networks or stored and retrieved to and from hard disks and other media, and protected in effective ways against corruption by channel noise. An interesting final remark in that sense is that the theoretical limit that Shannon established for the capacity to correct errors was approached rather slowly in the last half of the twentieth century, through the work of a multitude of researchers, and that today's best error-correcting codes work practically on that limit. Altogether, these achievements herald today's convergence of theoretical work with computer science and technology and clearly have a major impact on today's society.

1.4 Mathematics and e-Learning

Here we quote just a few points from the introductory part of [27]:

> The expectations created by e-Learning are certainly high, at all levels, and we may wonder how much of it is going

to be true, and up to what point can it help in the case of mathematics.

The reasons behind the high expectations on e-Learning stem from well-known characteristics of the e-Learning systems:

- In principle, access is possible from anywhere and at any time, thus making possible flexible (even just-for-me) and just-in-time courses of learning.
- The teacher can also be anywhere and do most of his teaching job at any time (preparing materials or following up and coaching his students).
- It allows for synchronous activities of a teacher and a group (at an agreed time), but again without restriction on the location of the people involved, and, what is more, with the possibility of addressing a much larger audience than a conventional class.
- Assessment can be automated to a large extent and final grading can be integrated seamlessly into the institution's information system.
- The learning materials and experiences can be richer in many ways, and they can be easily maintained and updated (as compared to preparing, say, a new edition of a paper book).
- There are also indications that e-Learning may induce deeper understanding and stronger retention.

For a general view of the main issues involved in e-Learning, see [146].

2 The Content Revolution

Technology is essential in teaching and learning mathematics; it influences the mathematics that is taught and enhances students' learning.

—USA National Council of Teachers of Mathematics

There is as much a need for the teacher to communicate mathematics to the students, as for students to communicate mathematics to the teacher. Since teaching methods will be increasingly measured by objectively proven learning outcomes in the students, and this through a variety of ways and circumstances, communication will increase in both directions.

It is thus desirable that the new learning environments have tools for the production and management of mathematical content that are available to everyone, and reasonably straightforward to use. Such tools should assist students in the task of formulating their required productions in the same way that they should help teachers with the job of encoding the learning materials.

How can this possibly be realized?

2.1 Encoding Mathematics

When doing mathematics on a computer, one has to take the view that mathematics is just like any ordinary language and, as such, it is meant to support communication. As Confucius said, "If language is not correct, then what is said is not what is meant. If what is said is not what is meant, then what ought to be done remains undone."

Any piece of computational software handles a well-defined proprietary internal representation of mathematics most suited for the kind of manipulations expected on the objects, be they complex numbers, equations, polynomial ideals, or formal proofs.

Computer-internal representations of mathematical entities often choose to employ the essential, necessary arguments required to identify an abstract object uniquely, for instance, a complex number $a + ib$ as the pair $\langle a, b\rangle$. The internal representation is chosen so that, being unambiguous, manipulations ("what ought to be done") can be carried out exactly and efficiently. However, computational software converts the internal representation to a more natural form whenever user intervention is required, usually by a "prettyprint" functionality.

While it is possible for a mathematician to interact with a computer program using a specific way to express mathematical objects, it is not so simple for different computer programs to exchange their internal representations. Clearly, the typeset expressions rendered at the user interface cannot be a candidate for unambiguous language, since different mathematical objects may be printed in the same way, e.g., a closed interval and a two-dimensional vector.

OpenMath [url:153] has been explicitly designed to support the communication of mathematics among different computer programs and to be pretty printed on a computer monitor via MathML presentation [url:183] or on paper via LaTeX.

Mathematical markup languages like OpenMath and MathML offer the possibility to represent mathematical content in a level of abstraction that is not dependent on localized information about notation and culture. This representation typically focuses on the semantics of the mathematical object and postpones localization aspects of mathematics to the rendering process of the markup. While the typesetting of mathematical markup has been the object of numerous efforts, from MathML presentation to SVG converters, the rendering of mathematics in a "verbalized" jargon has not yet received similar attention. The WebALT EU e-Content project has been devoted to the application of language technologies that automatically generate text from mathematical markup.

Mathematical jargon is an important aspect of the education of students. Not only does a teacher train pupils in problem solving skills, but she also makes sure that they acquire a proper way of expressing mathematical concepts. To our knowledge, digital e-Learning resources have used a representation in which text is intermixed with mathematical expressions, even in situations where the actual abstract representation, for instance of the statement of a theorem, can be reduced to a single mathematical object. One reason for this representation choice is that the rendering process would otherwise produce a symbolic, typeset mathematical formula that might prove too difficult to understand for the students or simply just too hard to read. However, by representing this kind of mathematical text in a language-independent format such as the one provided by markup languages, it is possible to apply language technologies and generate a "pretty printed" version that mixes text and symbolic notation to adapt to the native language and sophistication level of the reader. Languages covered so far include English, Spanish, Finnish, Swedish, French, and Italian.

2.2 Mathematical e-Content

Digital mathematics content, and the advanced tools for its creation, are among the main pillars in the communication of mathematics. We have to stress, however, that this is not as a straightforward as it may seem at first glance.

A commonly accepted estimate is that the cost of preparing high-quality interactive online learning materials is 200–300 hours of labor for one hour of student learning. A dedicated group of experts is needed to accomplish such a task. The preparation of a full one-semester course (56 hours) will typically require more than 10,000 hours of work by experts. This is months of labor for 100 people. With overheads, the total cost of the production of a 56-hour one-semester online course in mathematics amounts to about one million euros. This is prohibitively expensive. It is no wonder that, the teaching of sciences still mostly happens in the traditional way, which, in the long run, is even more expensive.

3 A Sample of Experiences

The wildness we all need to live, grow, and define ourselves is alive and well, and its glorious laws are all around.

—Robert B. Laughlin, *A Different Universe*

In this section, we will give a brief presentation of some of the experiences undergone within the WebALT project, or in closely related endeavors, that are related to communicating mathematics in the digital era. For more information on some of the topics, we refer the reader to [51] and [52].

3.1 WIRIS at Edu365

For our purposes here, it will be just enough to quote two paragraphs from [93]:

> Generically, WIRIS is an Internet platform which, on one hand, performs general mathematical computations solicited by its users and, on the other hand, supports the creation of Web-accessible interactive documents and materials.
> WIRIS is one of the main services offered by the Internet portal edu365 [url:9] of the Education Department of the Catalan Government.[2] Access to the site is unrestricted and only a standard Web browser is needed.

Users of WIRIS facilities are teachers and students both in secondary schools and in the universities.

3.2 Teaching and Learning Error-Correcting Codes

One of the earliest uses of WIRIS was to set up a web lab as a resource for the computational and practical aspects of the theory of error-correcting codes (see [276]). It contains over 140 entries, structured according to the table of contents of the paper book. These entries can be accessed by the links provided in the PDF book or directly from the webpage. In any case, the system can be used both by a teacher in a lecture hall equipped with an Internet connection and a

[2] At present, there are many similar WIRIS servers in several countries; see [url:192].

videoprojector, and by the students in the PC rooms. This model of digital content may promote autonomy in learning for some subjects, but in the coding theory class the benefit is limited to those who are really interested in the computational side of the mathematical subjects. The explanation of why the interest is not general is that, so far, the lab activities have not been taken directly into account in the final assessment.

3.3 Refreshing Mathematics at UPC

One of the needs at universities today, and especially in engineering schools, is to provide suitable means to help reinforce the mathematical background of first year students in topics like plane geometry, trigonometry, and basic facts about linear equations.

To solve these kinds of problems, we can only rely on suitable digital content and web labs. One possible design is the one followed by the EVAM project at UPC [url:58]. The virtual tools used in this project started with WIRIS technology integrated in Moodle and continued with the production of MapleTA exercises in connection with the WebALT project.

One advantage for users and for the academic community is that only a standard web browser with Java is required by the end user. In other words, users need no additional software. The interface and the computational engine can also be adjusted according to the specifications of the school involved. In any case, the architecture of WIRIS enhances the adjustment to the computer and the communications facilities available.

The methodology we have just sketched fits perfectly in the framework of the European Credit Transfer System. For professional engineers-to-be, this methodology helps students in gaining competences for working both individually and in teams, for managing time effectively, and for using computer resources appropriately.

4 The WebALT Content Architecture

Our course, which we called the Laboratory in Mathematical Experimentation *and which students called "the Lab," succeeded beyond any of our expectations.*

—G. Cobb *et al., Laboratories in Mathematical Experimentation*

The high cost of the production of online learning materials can be partly offset by preparing reusable materials. In the WebALT project, we have chosen to produce a large number of "educational lego pieces," i.e., modules that explain only one concept or method. Such lego pieces can then be combined in a variety of ways to assemble courses that serve different needs.

A module consists of the following components:

1. A short lecture ("Ten Minute Talk") that is a slideshow that can be used both in traditional contact instruction and in e-learning. A Ten Minute Talk may be a PowerPoint presentation or a slide show prepared by LaTeX (using, e.g., the Beamer class). For example, the various ways to define functions could be the topic of a module.
2. A set of solved problems pertinent to the Ten Minute Talk.
3. A set of unsolved problems.
4. Various laboratories that allow students to experiment with the mathematical concept at hand. Such laboratories may use a variety of techniques. In the WebALT Project, we have developed both WIRIS and Maplet laboratories and drills (java applets powered by the computer algebra system Maple).
5. A set of automatically graded problems administered to students by a system like Web Work or MapleTA that can automatically grade the students' responses.

Clearly, not all modules would contain all of the above. In fact, this architecture allows an incremental production of content and a continuous improvement of the quality. When teaching a course, for example, typically the first priority will be to have slides covering 1 and 2. Having a good set of unsolved problems is also part of the regular business of the content developer.

The really difficult parts, but also the most crucial ones for advanced learning environments, are those specified in 4, for producing good labs and question banks is time consuming and (for many) not easy at all, if only because doing so requires many different sorts of skills that generally cannot be found in a single person. In any case, laboratories have a great potential in education in general, because the applications that they enable render mathematics live. Maplets offer an alternative technology. Clearly, such laboratories can also be created using java. Systems like WIRIS make the creation of laboratories much faster and more fun.

4.1 Single Variable Calculus

This is an elaborate instance of this modular-structure organization [url:190]. The modules are accessed either through a category index or through an interactive labeled planar graphical tree whose leaves correspond to the modules and whose nodes correspond to category subdivisions of the contents. The navigation of the tree is easily accomplished by dragging nodes around.

5 Concluding Remarks

Who has seen the wind?
Neither you nor I:
But when the trees bow down their heads
The wind is passing by.

—Christina Rossetti, *Poems*

In conclusion, we stress the following points:

1. *Have clear aims.* Instruction has to be driven by aims and should be successful for most students. This is not currently the case. Overcoming this shortcoming must be regarded as one of the great challenges to today's educational systems, and we do not see a way out without widespread access to good technology-based learning environments.
2. *Diversity.* Learning environments should be able to cope with the great diversity of students in background, cognitive styles, attitudes, mental maps, and so on. This goal is not easy to reach, but again we think that this is the direction to take.
3. *Learning autonomy.* Educating students to become more and more autonomous, the learning environments should promote self-study and self-assessment, with as little learning overhead as possible.
4. *Assessments.* They will continue to be very important, since it is safe to say that students are basically driven by what will be required of them in the final assessments. This can be a good opportunity to enhance learning through a wide spectrum of self-assessment tests provided by the learning environment.

In the context of these desiderata, let us finish by pointing out some of the things that work and some of those that do not, as this gives an indication of what we should try to bring about and on what we can rely to that end.

Here are a few of the things that work reasonably well, and that can be expected to keep improving in the near future:

- computing and communications technologies,
- content formats,
- authoring tools,
- presentation tools.

And here are a few of the deficiencies that should be overcome:

- Suitable technology infrastructure in lecture halls, study rooms, and libraries is more the exception than the rule.
- The use of technology, when it is available, still represents a large knowledge overhead for teachers and students. Because of this, many simply refuse to spend the required effort. This resistance is often magnified by the lack of suitable academic rewards.
- Most teaching is still the same for all, delivered at the same time, and driven by subject matter—and not by the student's learning. This is a consequence of the preceding point.
- Use of technology, including present-day forms of content, tends to be dispersive for students. Overcoming this, by means of smart feedback systems, should be one of the most urgent concerns.

The IntBook Concept as an Adaptive Web Environment

A. Breda, T. Parreira, and E.M. Rocha

Any interactive multimedia educational software relies on a varied blend of multiple means such as text, graphics, animations, audio, and video, which are only achievable through the development of technological tools. In turn, it was the growth of interest in interactive open systems that led to new technological developments. A deep dynamic interaction between technology and interactive multimedia-based learning has been established. Interactive multimedia-based learning systems (ILSs) have evolved tremendously, beginning by focusing on self-directed learning with minimal or no teaching intervention, and progressing to highly structured, interactive collaborative learning environments. According to a cognitive theory of multimedia learning [178], a successful learning process includes:

- meaningful interaction with academic materials;
- selection of relevant verbal and nonverbal information;
- organization of information into corresponding mental models or representations;
- integration of new representations with existing knowledge.

As a target point, an ILS aims to maximize learning effectiveness, by means of interactive multimedia applications. Effectiveness of interactive learning systems is still an open and controversial issue. Until now, no theoretical base of evidence has been recognized. However, recent research studies on cognitive psychology have given some important guidelines (principles) for interactive instructional designers to create an ILS that is aesthetically appealing and pedagogically engaging. In Section 1, we shall describe these principles. According to Rachel S. Smith [237], to enable learning, we must have clear answers to the following questions: What educational involvement are we trying to explain? What do we want to achieve with the ILS we intend to develop? Every choice made during the design process must refer to the reasons that led us to start working in that specific project and to the underlying educational goals. Therefore, we must choose meaningful content that directly supports this goal and present it in appropriate ways. In other words, we need to mobilize syntactic and semantic elements, and use different symbolic systems (such as audio, still images, animation, video) chosen according to each situation [8].

1 Instructional Design Principles

The cognitive architecture for information processing is thought to consist of three components: *sensory memory*, *working memory*, and *long term memory*, corresponding, roughly, to the input, processing, and storage stages of computers. The flow of information begins through our senses (eyes and/or ears), being momentarily stored in the visual and auditory sensory system (input stage). It is then processed in the working memory, the center of all active thinking activities (processing stage). Despite its powerful ability to process information, the volume of information and the duration of the process are limited. The information is then stored in the long-term memory (storage stage). In the working memory, the information is temporarily stored, and it is involved in the selection, initiation, and termination of information-processing functions such as encoding data (integration of new knowledge from the working memory into long-term memory) and retrieving data (extracting the new information from the long-term memory and transferring it to the working memory to apply it within real-life context and applications).

The small capacity of the working memory is rather intriguing. Whereas the average number of items (words or numbers) for adults is seven, the capacity of long-term memory is very big (no upper bound has been established). Figure 1 illustrates the human cognitive architecture processing.

Schema formation and automation are two processes that help to overcome the limitations of working memory. Schema formations are cognitive constructs that incorporate multiple elements of information into a single element with a specific function. They are stored in long-term memory and can be brought to working memory. The automation of those schema so that they can be processed unconsciously dimin-

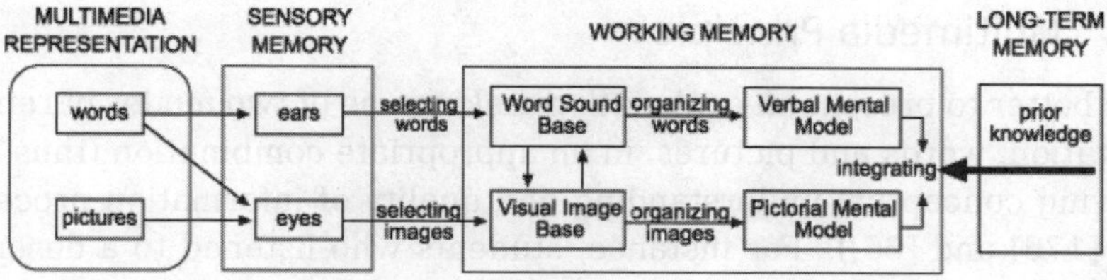

Figure 1. Representation of a cognitive theory of multimedia learning [191].

ishes the load on working memory. The 1980s saw the emergence of cognitive load theory (CLT) as a major theory providing a framework for investigations into cognitive processes and instructional design. CLT is concerned with instructional implications of the interaction between structures and cognitive architecture [201]. We may classify the cognitive load in three categories: *intrinsic cognitive load, extraneous or ineffective cognitive load,* and *germane or effective cognitive load.* Intrinsic cognitive load leads with the demands on working memory capacity imposed by element interactivity (intrinsic) to the material to be learned. The element interactivity is intrinsic to the material and so cannot be altered by instructional manipulations. Extraneous or ineffective cognitive load is the unnecessary load imposed by instructional procedures. It is determined by the interaction between the nature of the materials being learned and the level of expertise of the learner [183]. A great part of the work in cognitive load theory is concerned with procedures for reducing ineffective cognitive load. Germane or effective cognitive load is also influenced by instructional procedures, but whereas extraneous cognitive load obstructs learning, germane cognitive load enhances learning. Germane cognitive load refers to the working memory resources devoted to schema acquisition and automation; it is associated with processes that are directly relevant to learning. Cognitive load theorists believe that intrinsic, extraneous, and germane cognitive loads are additive, in the sense that, together, the total load cannot exceed the working memory resources available if learning is to occur [201]. Their main instructional principle is, therefore, to decrease extraneous cognitive load and increase germane cognitive loads within the working memory capacity. Next we shall enumerate principles for instructional design makers, keeping in mind the cognitive load theory.

1.1 Multimedia Principle

It is better to present an explanation making use of two modes of representation, words and pictures, in an appropriate combination (thus improving conceptual understanding and quality of information processing [176] and [66]). For instance, students who listened to a description of a composition of isometries while also viewing the corresponding animation had a better understanding than those who listened to the narrative without viewing the animation. Users build two mental representations (visual and verbal) and construct connections between them.

1.2 Contiguity Principle

An explanation containing words and pictures is better understood when they are presented at the same time. Since corresponding words and pictures are in the working memory at the same time, schema formation may take place [176].

1.3 Modality (Split-Attention) Principle

When giving a multimedia explanation, avoid splitting the user's attention by using multiple sources of mutually referring information [176, 177]. The modality principle asserts that verbal information is better understood when it is presented in a narration instead of in visual on-screen text. It is a way to prevent an overload of visual information [66, 191].

1.4 Redundancy Principle

This principle states that users learn better from animation and narration than from animation, narration, and text if the visual information is presented simultaneously with the verbal information. Multiple representation increases extraneous cognitive load [66, 191].

1.5 Spatial Contiguity Principle

Users learn better when on-screen text and animations are physically integrated rather than separated [237].

1.6 Individual Differences Principle

This principal is based on the assumption that multimedia, contiguity, and split-attention effects depend on the user's personal characteristics. Users who lack prior knowledge tend to show stronger effects than the ones who possess high levels of prior knowledge [176]. Allowances for varied learning styles should be integrated into the instructional design.

1.7 Learner Dynamics Principle

A conceptual distinction should be made between applications that are essentially content delivery (learners progress through the educational materials in a traditional way) and applications containing interactive exploration (highly interactive, with simulations, games, etc.). Simulations can provide goal-based challenges that stimulate user motivation and interest in the material being presented. As stated by [133], providing tools for annotation and collation of notes can be effective in stimulating learner engagement.

1.8 Mediation Principle

Another aspect that has a profound effect on the way that learning occurs is related to the structure of the activities (tasks) proposed. There is no question about the importance of the student feeling engaged with the activities and getting actively involved in the learning process. The ITS must give to the learner the opportunity to solve problems, draw conclusions, compare options, and think about what he/she is doing. Accordingly, the range of activities should address different modalities of learning to contribute to the development of his/her synthesis, analysis, and evaluation skills, never forgetting to give the student immediate feedback on his/her progress. The interaction must happen not only between the student and the system, but also between the student and the educator or between the student and other students, in order to facilitate the comprehension and manipulation of the information. This, as stated by [8], is an issue we cannot forget while designing a collaborative ITS.

1.9 Coherence Principle

Students learn better when extraneous material is excluded from, rather than included in, multimedia explanations [176, 191].

2 Graphic Design Principles

Finding the balance between the mathematical content and an enjoyable learning experience for the students is not easy. To help in this stage of the development, Rachel S. Smith [237] presents a group of guidelines to keep in consideration when designing the learner's experience in terms of graphic design and usability.

2.1 Balanced Design Principle

The graphic design of the interface involves a lot of choices regarding the layout, colors, navigational elements, and user controls, as it defines the way the student will access the content. First of all, the page layout should be visually balanced. The goal is to distribute the visual weight uniformly but at the same time to avoid perfect symmetry and center alignment, because even though this is safe, it becomes visually boring. Colors and layout should be similar from page to page so that the student understands the navigation and remains interested in the book. But for different sections, e.g., the exercise pages, we must design a contrasting layout, so that the student can tell where he is and what he is expected to do.

2.2 Interest and Continuity Principle

Even though it is widely recommended to keep a clear and simple design throughout the book, Rachel S. Smith stresses that the existence of an element repeated on every page adds interest and continuity, like a little animal that appears in every illustration of a children's book: it is encouraging and stimulating. This character is the engaging agent that gives rhythm to the action and at the same gives feedback regarding the student's performance. Therefore, including animated agents in an interactive ILS could be a promising option in the interface development because they allow us to use communication styles with which the user is already familiar. An important aspect in the creation of these agents is to look for adequate audiovisual appearance. Also, the success of an agent in terms of student acceptance and interface efficiency depends much on its communication skills and behavior in general (body language, facial expressions, vocabulary, locomotion, and interaction with other elements of the scenery). It is important to point out that characters must be created as distinct individuals with their own knowledge, interests, personality, emotions, and appearance that will influence their actions [10]. In this area, the design team gives reins to its imagination and looks for creative solutions that combine reality with the unreal, creating a universe of its own with virtual objects. Although these objects do not necessarily have a real equivalent, the consistency of their representation lets the user interact with them as if they were real. In the end, if we take these guidelines into account and create a graphically simple and coherent design, combined with some rhythm and contrast, the elements should work together as a whole. If we want to catch the attention of the student more easily, the language we use is the clue. In fact, rhetoric is an expressive element capable of producing the desired effect in the user. Using a suggestive and sometimes risky language intends to persuade the user and direct his actions. Humor is a good way to cause empathy, and therefore to hold the user's interest. In addition, if we introduce the surprise effect, by changing the logical sequence of the speech and adding unexpected situations, it refreshes the speech and increases its dynamism, relieving some tension for the user. In this case, illustration and animation will assume a leading role during the entire book, by introducing a little nonsense associations and emphasizing some elements [208]. Metaphors are also a good option, but must be chosen carefully. Good metaphors generate visible images in the mind so that we can bring them alive by appealing to the student's perception (sight, sound, touch) and stimulating his/her memories. They suggest the familiar and at the same time add a new twist to the action [259].

2.3 Usability Principle

But how easy is it to use an ITS? Can the user easily find the information he/she needs? Are instructions clear and easy to understand or is the interface so distracting that the user has difficulty focusing on the content? Knowing the answers to these types of questions lets us know the usability level of the ITS. First of all, we cannot forget that users must spend their time enjoying learning with ITS, rather than figuring out how to use it. If we create the interface as simple and clear as we can, we will surely improve the learning experience. As we have already observed, the first step in doing is to create a consistent appearance of the pages in terms of layout, colors, and overall look. Then, all the navigational elements must look similar and should be placed in a clearly identified area. And, above all, we must be able to anticipate the student's wants and needs. If the ITS is easy to understand and to interact with, the student will learn how to use it quickly, and hopefully will make fewer mistakes. Another key aspect is to give the student the feeling that he is in control of the book. If all the pages provide a clear navigation and allow the student to stop, restart, skip, or revisit animations and videos, it is more likely that he will explore the book because he knows he will not feel lost. Moreover, the student should never feel trapped; he should have a clear way out.

3 IntBooks

Technically speaking, we can define IntBooks as a web technologies aggregation platform [url:117]. Thus, the aim is to build a document platform that aggregates (as many as possible) web technologies, giving the reader an interactive, intelligent, and rich (web) environment, and the author a "simple" way to build and share it.

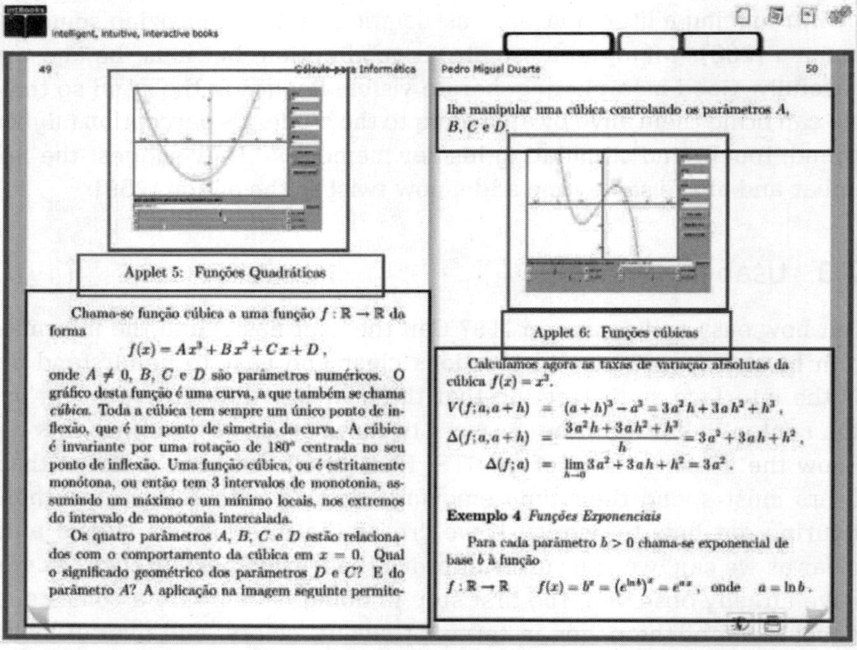

Figure 2. A fullscreen snapshot of a calculus IntBook, where different source fragments are marked with boxes, i.e., LaTeX(the bottom boxes), Java applets (the interactive graphs), and HTML (the figure captions).

3.1 Language and Structure

In order to get the desired aggregation of the different web technologies, we define a specific language (file format) designated "IBK," specified in the XML format. A book is represented by an IBK package—that is, an IBK file—containing all the structural information needed to build an IntBook, together with a set of resources (external files). The IBK file (the main file) is composed of a header section and a body section. The header section contains metadata about the file, authors, book's layout (single/double page), language settings (default and translating languages), list of resources, and tuning options, such as $\LaTeX$ compile parameters. The body section contains the book content — text and objects/components that will dynamically generate the set of pages. The main text allows the full use of $\LaTeX$ and may have embedded objects that are building blocks of a page (Flash, Java Applets, HTML, etc.). Objects are embedded in the body section using simple IntBooks commands, which greatly simplify the build process for authors, so that they do not need to be aware of the technology used. In the future, this may also allow authors to incorporate seamless blocks of text (as theorems) in their texts, available in some kind of "text pieces" library.

3.2 Book Processing

The main component of IntBooks is the "pipes processing" mechanism, which allows IBK files to be sequentially transformed, so that their content is converted between different formats. Figure 2 shows a snapshot of an IntBook in which the IBK source has some $\LaTeX$ fragments. However, the final versions of such fragments (seen on browsers by readers) are presented as Flash objects. In fact, Flash is a viable alternative to PDF and MathML for presenting mathematics on the web.

The book's processing itself is a complex task, which, in order to be time efficient, must be done in a cluster of computers, e.g., using openSSI technology (see Figure 3). The first step is mainly validation, the second step is the pipes processing (according to each book), and the final step is the call of a web application (C#). Here, the book metadata is added to the system SQL database and all the static pages (for the several languages) of the book are generated. After this, the book is ready to be consulted by users that have the necessary credentials.

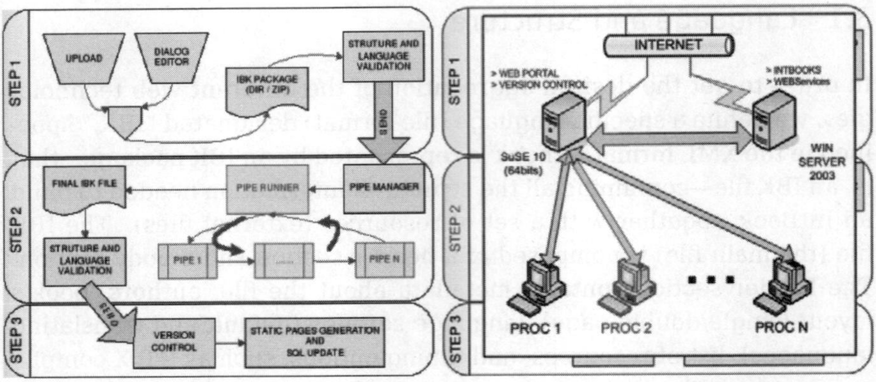

Figure 3. The book processing and IntBooks architecture.

3.3 Book's Navigation

The user's navigation in an IntBook is defined by an XML file, which contains information about the content structure. Usually, the navigation file has a pure sequential form (linear structure), and is the same for all users of a given book. On use, a personal navigation file is created for each user in order to adapt the book to the user's answers to the book's questions (exercises) and their last seen pages. In this case, each user will have a personal navigation model (nonlinear structure). An improvement can be made by using Expert Systems (artificial intelligence systems), with a knowledge base and a well-defined set of production rules, in order to generate a best-suited personal navigation file. Although this mechanism is not yet implemented, it is a strong possibility for future work.

4 TexMat: A Mathematical IntBook

TexMat is an interactive mathematical IntBook planned for the fifth and sixth grades of Basic Portuguese School (ten to twelve year old youths). In fact, TexMat was the application that originated IntBooks. Instead of developing a system controlling a specific ITS, we have expanded the idea for a more general and abstract system that can build and control every book designed to be an IntBook, where TexMat can be included. Even so, TexMat has some additional features that are not present in all IntBooks, such as the Notebook and Agent applications.

4.1 TexMat Instructional and Graphic Design

TexMat's contents follow the Mathematics National Curriculum of the Portuguese Basic Education. In its design, the instructional and graphic design principles have been taken into consideration. In the development of the geometry units, we have followed Duval's theoretical framework [89] with respect to the cognitive processes involved in geometrical reasoning. Next, we illustrate some of the instructional and graphic principles present in TexMat.

Layout. TexMat, following the Mathematics National Curriculum of the Portuguese Basic Education, is divided into four study units: geometry, numbers and calculus, statistics, and direct proportionality. We have associated to each of these units a different color. The layout of each unit and subunit is similar in terms of shapes, fonts, and sizes, with the color scheme being the distinguishable element. In this example, for the geometry unit we chose pink as the base color to be used in type fonts, exercise sheet, and separators. TexMat presents a structure similar to a book, where the navigation is made by turning pages and every concept is presented on a new page, exactly the same way as in traditional didactic materials. Inside, each page has a vertical area for the contents, which remains identical in the entire book. Each page also has a clear buttons area. On the right side are the buttons to access the learning help tools (glossary, notebook, and evaluation sheets), and on the top are the ones that link the user to the different units. On the left side are the buttons that operate on the current page ("add to the notebook," "print," "sound ON/OFF," and "quit"). The contents are explained to the student textually, next to different and easily identifiable areas where the animations and video that illustrate the text appear.

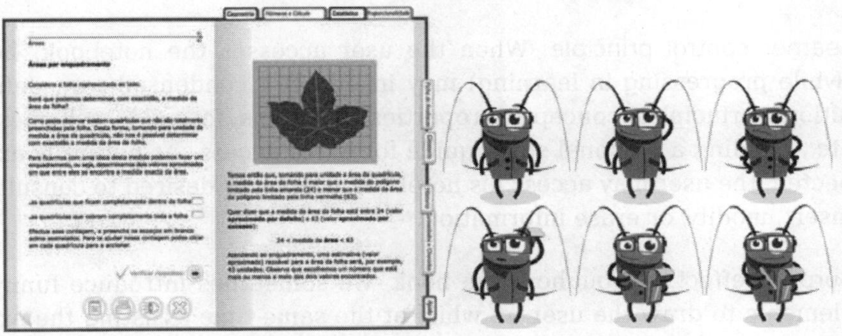

Figure 4. TexMat page example and animated agent.

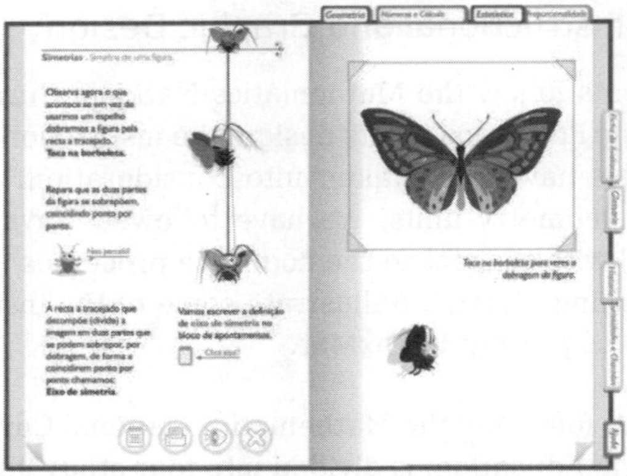

Figure 5. When the learner touches the butterfly, an animation shows its axis of symmetry.

Animated agents. TexMat books appear inhabited by four characters—bookworms, the agents that will guide the student through each thematic unit. They will give feedback to the student's actions and help him if he ever feels lost. Each agent has its own personality and a different occupation that represent the content's applications to real life situations. For example, in the Geometry Unit, the character appears as an architect.

Contiguity principle. Several explanations simultaneously contain text and (animated) pictures (see Figure 5).

Learner dynamics principle. The learner deduces an expression for the area of a circle of radius r, simulating the division of the circle into equal parts (see Figure 6).

Learner control principle. When the user accesses the notebook, he (while progressing in learning) may insert, in a condensed way, definitions, principles, concepts, properties, theorems, formulas, schemes, etc., building a personal study guide for quick access. As naturally expected, the user may access his notebook whenever desired to consult, insert, modify, or erase information.

Surprise effect. Throughout the book, we sometimes introduce funny elements to draw the user in, while at the same time assisting the explanation. Several examples appear with many illustrations and animations that are used to increase the dynamism and to refresh the speech.

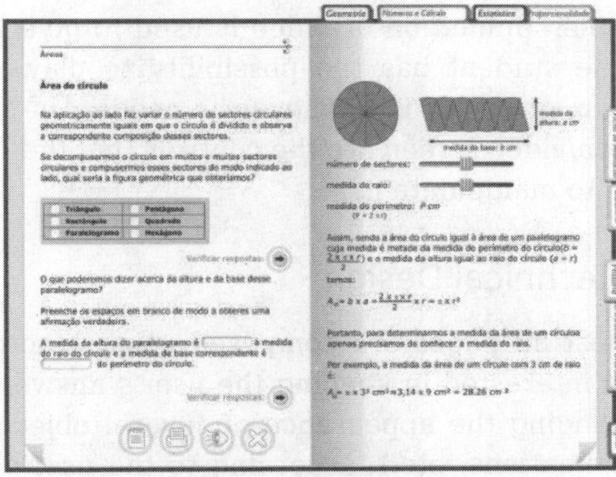

Figure 6. The learner is invited to use the application to simulate the division of the circle into equal parts and try it with a different radius. By doing this, he finds out that to calculate the area of the circle he needs to know its radius. He then deduces the expression for the area of a circle of radius r.

Activities structure. Activities are randomly generated and categorized in three classes: reproduction (problems whose solution is a direct application of concepts and procedures), connection (problems requiring effortless routines involving distinct concepts), and reflection (nontrivial problems involving more than one concept). The exercise pages differ from the explanation pages, and the layout of each of the three different classes is in the same colors going from lighter (class 1) to darker (class 3), see Figure 7. The activity range includes matching exercises, filling spaces, true/false, drawing, constructing figures, and manipulating elements.

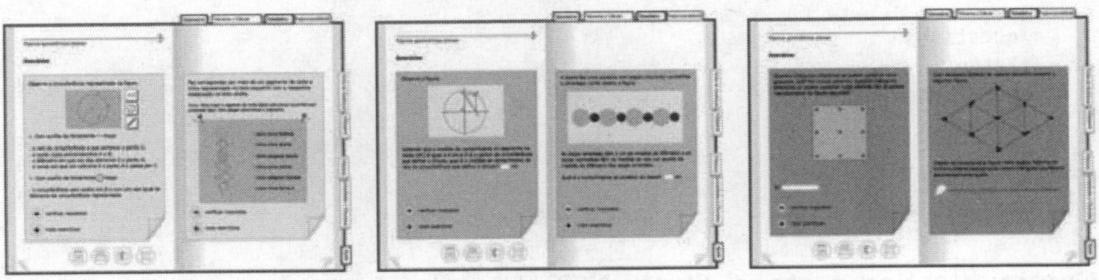

Figure 7. Examples of generated activities with increasing difficulty, from left to right.

Control keys. When animation or video is used to give explanations, the student has the possibility to play, stop, restart, skip, or revisit it whenever he needs. Under the animation/video screen are the controls that the student can use to manipulate it.

4.2 TexMat Technical Design

In TexMat, almost all pages are composed using Macromedia Flash. Because we are interested in keeping the user's answers to the questions and in changing the appearance of several objects (visibility of text, images, animations, etc.) according to the user's accesses and interactions with the book, it was necessary to define a communication protocol (using XML) to exchange information between the Flash objects and the database, which will be presented in continuation. In the following are some parts of the TexMat IBK file.

```
<intbook-system>
<book id="003" title="TexMat" authors="Geometrix" pubkey="...">
 <header>
  <lang code="pt" />
  <translate-lang code="en" />
  <layout name="texMat"  menu="yes" style="texMatStyle">
    <set button="b01" pageid="43" /><set button="b02" pageid="91" />
  </layout>
 </header>
 <body>
  ...
  <sheet id="52" nid="53" lid="51" comesfrom="body"><page>
   <flash filters="user">
    <vars_layout fName="GF20" baseSrc="flashlets" w="712" h="476" />
    <vars_state>
      <var id="17344" name="text11" property="_visible" value="0" />
      ...
      <var id="17348" name="select2" property="image" value="cr2.swf"/>
    </vars_state>
    <questions>
      <question id="2617" answer="" attempts ="3" />
      ...
    </questions>
   </flash>
  </page></sheet>
  ...
 </body>
</book>
</intbook-system>
```

Communication protocol Flash-.NET. In order to exchange information (dynamic variables, users' answers, etc.) between Flash objects and the server, we defined an XML protocol to communicate between each Flash file and a .NET application, which interacts with the SQL database. When each Flash file starts, a request is made to the getSetVars service, with information about the block's identifier (received by flash-vars), with "action=getXml," for the XML referent to that particular instance. The Flash file gets and manages the XML, and actualizes the values of the variables and the questions. This appends exactly when the Flash is loaded. When Flash finishes, that is, when another book's page is called, a javascript function tells the Flash object to move to a specific frame. Here, all the changes that happened relating to the dynamic variables and the questions are actualized in an XML file with a similar structure to that received, which will be sent back to the get-SetVars service, with the information about the block and, now, with "action=setVars." The getSetVars service (developed in .NET) is used to communicate between the database and the Flash objects. When it is called with the action "getXml," an XML transference file is written, by consulting the database, with the user data and the relative block and is sent for Flash's reading. When it is called with action "setVars," the received XML is read, and the new values of the dynamic variables and the questions are put (or actualized) in the database.

Notebook and agent applications. The notebook and agent applications are developed in ASP.NET, and are located in the same project that controls the book. The notebook application is called by the user at any moment he wants. If it is called from inside the Flash object, it goes directly to the respective definition of the associated concept. Otherwise, the notebook application opens in search mode, permitting the user to search every concept in his own notebook. Each concept is are automatically added to the user's notebook when he walks by the page that contains the concept's explanation. This relation between the concepts and the pages of the book is also defined using XML. In each concept in

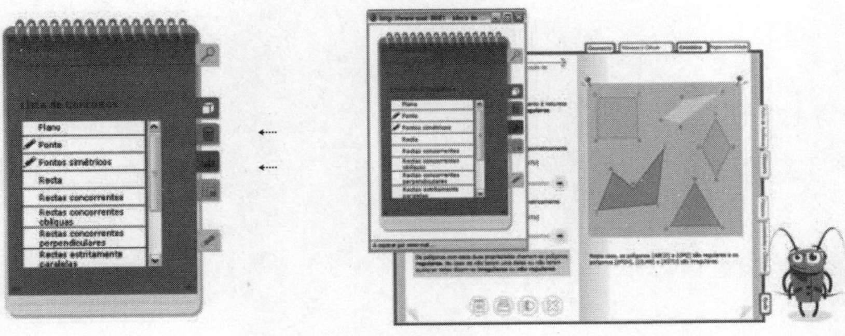

Figure 8. Notebook and animated agent.

the notebook, the user can add a personal note which will be kept in the database. The next access to the concept's definition will contain the personal note added, which can be edited or deleted at any time. Notebook also permits personal notes that are not necessarily related to the book. The Agent application manages an interactive agent through the display, during the user's navigation through the book. It controls the expressions and the movement of the animated GIF in agreement with Flash objects, instructions (related to the user's answers) and with the user's navigation. This application is still in development.

5 Conclusions

The use of animation as a learning vehicle can be very successful, as it takes advantage of immediate language, rich illustration, and fantasy to draw the student in, and to promote attitudes different from daily routines. The sensations and collective emotions that the audiovisual provokes in the students have a big pedagogical impact when appropriately explored. Animation makes the message clearer and easier to understand, and after they receive it, the students will translate and assimilate its content. The choice and the organization of the different elements of an ITS must keep in consideration the way that we are trying to approach the student. It is truly recommended to use real-world examples whenever possible so that the student can make a connection between the learned contents and the student's own life. The information must be demonstrated in an impressive way and, if possible, using concrete examples rather than abstract ones. However, it is also important to present complex scenarios that provoke a multiplicity of opinions and invite the student to think.

An Educational Environment Based on Ontology

K. Sugita, T. Goto, T. Yaku, and K. Tsuchida

In this chapter, we present a proposition for a Document Type Definition (DTD) for electronic versions of mathematics textbooks, a Resource Description Framework (RDF) for mathematical subjects based on mathematical text documents, and a supporting system for the self-study of mathematics. These are based on the characteristic structure of textbooks that are actually used in classrooms, mathematics dictionaries, and the concept of ontology. These are extensions of various documentation systems, the RDF system, and DC/LOM abstract models for metadata. We also construct an educational environment using these items.

1 Introduction

Students have a lot of mathematical content to learn, but they also need to share their school time due to the increasing demands of other subjects such as science, language, and social studies. As a result, curriculum is divided into fragments lacking logical unity, teachers tend to put emphasis on repetition of exercises at an elementary level, and students lose sufficient time to consider problems. These facts are believed to prevent students from understanding mathematics.

In order to deal with this problem, some websites are trying to offer good information about learning mathematics [1, 97, 256]. However, students need some background knowledge to utilize these open resources for their study. For example, a student may find a website that shows a method to draw a regular pentagon, but the site probably does not teach him the relation between regular pentagons, de Moivre's formula and other regular polygons. Also, students will have difficulty finding websites that supply good exercises to practice their recently acquired knowledge.

Ideally, students would have a perfect textbook that is logically composed and useful as both a learning tool and as a dictionary. However, meeting these goals results in a big textbook and raises another difficult problem for beginners.

Considering these situations, a supporting system will work well to bridge these gaps. Alternating a contents table, index of books, and dictionaries, RDF helps students search for objects within some resources, and searching systems using RDF are becoming dominant in various regions [32].

There are some excellent document-processing systems such as the DocBook [194] system that is used for general purpose XML documents, and LaTeX systems for documents containing mathematical expressions. However, using these documents as resources requires a lot of manual work to construct a support system.

We present a system as a solution to this problem that includes a preparation of resource documentations through construction of RDF and a user interface that provides both self-learning for students and an overview of curriculum for teachers. This work is inspired by the structure of textbooks that are actually used in classrooms and the mathematical dictionary edited by the Mathematical Society of Japan [198].

2 Mathtext.DTD

2.1 Characteristics of Mathematics Articles and Textbooks

In reviewing various textbooks and articles on mathematics, we observe that these documents are composed of the presentation elements that are set forth below. These elements guide students in learning the content and show relationships with other content. These components are called *learning attributes* (see Figure 1), and a document that contains learning attributes and obeys proper syntactical rules is called a *Mathtext* (see Figure 2).

There is an analogy between the structure of a textbook and a chart of the universe. Letting the universe of mathematics correspond to the general universe, a textbook corresponds to a chart of the universe, a content or a subject of mathematics corresponds to a part of the uni-

```
Learning attribute = { introduction, background, preparatory, assumption,
                       definition, formula, algorithm, proposition,
                       theorem, proof, demonstration, illustration,
                       figure, graph, animation, example, application,
                       summary, exercise, solution, hint, progress,
                       remark, short_note }
```

Figure 1. Set of learning attributes.

verse, and learning attributes correspond to guideposts in a chart such as points, lighthouses, and landmarks. A learning attribute is generally presented by a title of a section that contains the content, and presenting a logical sequence of learning attributes is an important factor in textbooks.

2.2 Syntactical Definition of a Textbook

Next we consider a logical definition of a textbook. It is the same as a definition of general documents and textbooks, so elements defined in other XML documentation systems such as DocBook may be applied to our definition. One important merit advantage of having a formal definition is that authors are able to get a uniform rendering of documents through XSLT without being bothered about it. The list in Figure 2 sets forth the minimal elements of the Mathtext syntactical definition. As a block element, `<paragraph>` seems to be the smallest block in normal XML usage, so we create a super block of `<division>` for syntactical elements, such as a learning unit.

2.3 A DTD for Mathtext

We next define the basic semantic parts of Mathtext. These are placed as attributes in syntactical tags in Figure 2, and appear as a `content` value. Our principle definitions are as follows, described within the context of maintaining a tree structure:

- `<chapter>` or `<section>` are root-levels of a learning content, and have requested `content` and `prerequisite` attributes, for example, `<section content="trigonometric function">`. These attribute value work as property values of *is-a* or *has-a* relations in RDF (which are discussed in Section 3.3). Generally, the `content` value is the same as the `<title>` phrase.
- `<division>` is a node to block elements, display elements, and lists, and represents a learning attribute of its block for the section containing it. Its learning attribute in Figure 1 is indicated as

Root <mathtext>.

Components <text_info>, <subject_info>, <preface>, <table_contents>, <chapter>, <appendix>, <reference>, <glossary>, <table_formula>, <table_function>, <bibliography>.

<chapter> includes a sequence of <preface>, <section>, <appendix>, and <reference>.

Sections <section>, <subsection>, <sub-subsection>, ...

Division <division>.

<division> is included in sections, and includes block elements, display elements, and lists.

Block Elements <paragraph>, display elements, lists.

Display Elements <equation>, <figure>, <graph>, <illustration>, <animation>, <table>, <quotation>.

Lists <itemizedList>, <numberedList>, <variableList>, <listItem>.

In-line elements <emph>, <eq>(in-line mathematical expression), <qt>(in-line quotation).

Cross references <cite>, <ref>, <link>, <preparatory>.

Miscellaneous elements <title>.

<title> is used to identify the title of documents, part of documents, and captions.

<equation> and <eq> will be processed by MathML [186] in the near future.

Figure 2. Elements of Mathtext.

an attribute value of att, for example, <division att="proposition">. This value is shown as a head mark of its block automatically via XLS processing, and also works as a property value of itself in RDF.

- The elements "figure," "graph," "illustration," and "animation" in Figure 1 are changed to syntactical elements <figure>, <graph>, <illustration> and <animation> respectively. Namely, <division att="figure"> is changed to <figure>, etc.
- Generally, the value in <preparatory> is most closely related internally to its chapter or document. On the other hand, the prerequisite value in <chapter> indicates a more global relation beyond its chapter or document.
- <division att="exercise" grade="A"> indicates the grade of this exercise, and its value is shown in subtitles.

2.4 An Example

Case Study. In Japan, school textbooks are required to be based on the government guideline for teaching. This guideline is worked out by the committee organized by members of specialists from each subject field and education. Appendix C shows an extract from the guideline about trigonometry in mathematics.

As one can see from this extract, one topic is divided into many years and classes. This is done in order to correspond to varying scholastic abilities of students, and future courses taken by students, and is called the spiral methodology of education. However, this division obstructs unified learning of mathematics, which is one reason to construct a supporting system for self-study of mathematics.

An example of Mathtext. In Appendix C, you will find an example of a document based on the Mathtext.DTD. This is a textbook of Mathematics II corresponding to the guideline shown in Appendix C. There, we show only a tag sequence omitting statements. As you may see, the content of the `<subsection>` tag is very short, mainly because that one subsection corresponds to one school hour.

3 Mathtext.RDF

3.1 RDF and Metadata

In the semantic web, one ideal approach is for an ontology dictionary to be generated from open resources themselves [15, 68, 83, 175]. But as mentioned in the introduction, it need not rely on open resources in our case. There is another approach concerned with IT education [155], but this approach treats only dependency or learning order among learning content and does not work for student learning. There is another World Wide Web approach that defines metadata about abstract information of resources such as DC (Dublin Core) [74], and it works as a retrieval index of resources. There are promotions to utilize the RDF concept for educational resources using these metadata such as DCMI (Dublin Core Metadata Initiative) [75] and LOM(Learning Object Metadata) [139]. These metadata are also useful in our case to identify a textbook as itself, so our proposal may possibly be treated as an extension to these metadata. For example, DC/LOM contains elements labeled "Coverage," "Description," "Relation," and "Subject"—these elements could indicate the contents of Mathtext resources and allow links between Mathtext documents and other learning resources.

3.2 Universe of Mathtext RDF

There are two basic strategies used to construct a learning object database. One is a database of resource identification information such as DC/LOM, and the second is a big knowledge base of learning content based on textbooks, dictionaries, and other open resources. However, in putting together many resources, there may be a contradiction in dependencies between various subjects. In order to avoid this problem, the database tends to become a simple dictionary type—one without relation and discipline. Considering these situations, we restrict the scope of RDF to only one textbook or one dictionary, and name it Mathtext.RDF. In this concept, one textbook or dictionary makes one universe and there is one Mathtext.RDF for each textbook or dictionary, and each universe is related to other universes using DC/LOM value of <subject_info>, <text_info>, attributes. Some differences between other RDFs are expected to inspire students to understand the situation.

3.3 Metadata of Mathtext.RDF

RDF is a set of triples (resource, property, value), (subject, verb, object), or (object1, relation, object2). This triple is interpreted such that a "resource" has a "value" about "property," that "subject-verb-object" or that there is an ordered "relation" between "object1" and "object2." If the subject of the resources is one such as DC/LOM, a set of properties and their values is called a metadata of a resource. In our case, RDF triple is defined as the following:

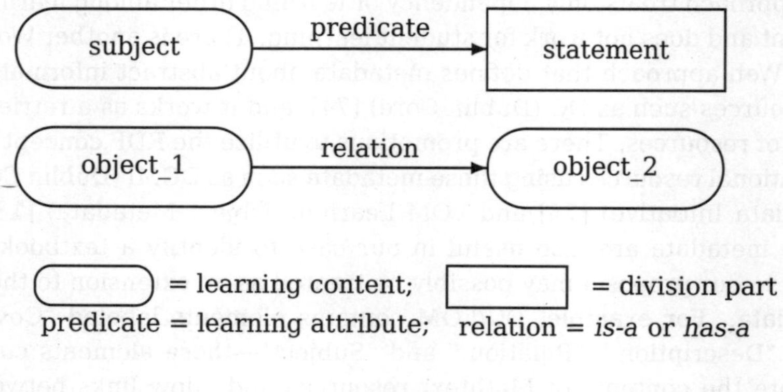

Figure 3. Image of RDF Triple in Mathtext.RDF.

Resource or subject. Unlike other RDF such as DC/LOM, in our definition, a resource or subject of Mathtext.RDF is one of mathematical content such as logic, quadratic equations, trigonometry, complex numbers, systems of linear equation, differentiation, and so on.

Property and its value. Two kinds of properties are provided. The first is a dependency relation between learning contents. This is also classified into two elements, *is-a* and *has-a* relations. The second is a property of a learning attribute shown in Figure 1.

- The value of *is-a* comes from a `prerequisite` value in a `<chapter>` or `<section>` tag, or `value` in a `<preparatory>` tag. However, there is a serious problem. That is to say, the URI corresponding to *is-a* value could be in another resource outside the Mathtext document. This situation causes an uncertainty in the selection of an *is-a* target.
- The value of *has-a* comes from the value of a progress tag. However, it is difficult for authors to describe detailed *has-a* content relations in the chapters and sections. Learning content contained in each chapter, section, and subsection is described in the `content` value, so *has-a* relations are able to be extracted from the `content` values of each `<chapter>`, `<section>`, and `<subsection>` corresponding to the syntactical tree structure of a Mathtext document as the LaTeXsystem generates a table of contents.
- The values of learning attributes come from the `att` value of `<division>` tags. These are defined in Figure 1.

Theoretically, *is-a* and *has-a* relations are inverses of each other. However, the authors may not pay much attention to this progressive area. Therefore, *is-a* and *has-a* relations are restricted locally in Mathtext documents.

We note that the information of one Mathtext document constructs an entity relationship diagram of a database. Here, *is-a* or *has-a* relations correspond to relationships, and learning attributes correspond to attributes of an entity. Extending this database to other resources, difficult problems arise concerning dependency, priority sequence, and consistency of learning content.

3.4 Generation of Mathtext.RDF

Mathtext.RDF is automatically generated from Mathtext documents using XSLT. However, as mentioned above, a resource corresponding to

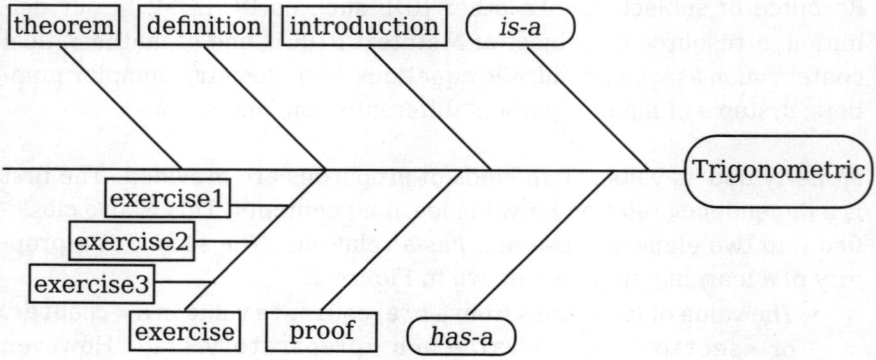

Figure 4. An example of a fish-bone diagram

an *is-a* value could be in another resource, and in that case, there is no algorithm to establish the *is-a* relation. One possible solution is to link to a universe composed in the same series of textbooks or works cited in the bibliography. But this requires manual intervention, and we are not sure if this is the correct answer. If one intends to extend a universe to open resources, there will be more complicated problems and a need for a more sophisticated ontology framework.

4 Application

There are some tools for representing RDF, but there is not a standard tool for it. Some utility programs need auxiliary information to form an RDF diagram. These situations show a widely spread application area of RDF.

In Mathtext.RDF, two kinds of applications are possible. One is a dependency diagram about learning content. This application uses only *is-a* and *has-a* property values, and is useful for teachers and curriculum developers. Second is a supporting system for the self-study of mathematics. This application uses *is-a* properties, *has-a* properties, and learning attributes. For this application, a "fish-bone" diagram is used for expressing Mathtext.RDF information.

A fish-bone diagram, or a cause-effect diagram, was originally used by manufacturers for analyzing causes of observed defects in the area of quality control. In the area of manufacturing, the building of a cause-effect diagram starts with identifying an effect whose causes one wishes to understand. To identify the causes, first some major categories are established. Using the effect and major causes, the main structure of a diagram is made—the effect as a box on the right side connected by a straight horizontal line, plus an angular line for each major cause connecting to the main line. For analyzing the causes, the key is to continuously ask the question "why does this cause produce this effect?" The answers to these questions become subcauses and are presented as short horizontal lines connecting to the line for a major cause in the diagram. When all the causes are identified and marked in the diagram, the final picture looks like a fish-bone structure and hence a cause-effect diagram is also called a fish-bone diagram.

In the case of Mathtext, at the head (effect) part of the fish-bone diagram, the learning content about which a student wishes to study is placed, *is-a* properties are placed at the end of each angular line (cause part), and *has-a* properties and learning attributes are placed according to the RDF. Each value is shown with its URI. If there are many learning attribute values such as exercises, these values are placed at the short horizontal (subcause) lines. Reversing the direction of RDF triples in Figure 3 and gathering RDF triples with the same subject (learning content) and same learning attributes together, one can get a fish-bone diagram of the content.

The user is able to select a dependency diagram or a fish-bone diagram. Clicking a node that a student wants to explore on a fish-bone diagram, the statements, illustrations, graphs, formulas, tables, and animations, if possible, that are contained in a selected division are displayed. For example, if a student selects a fish-bone system and enters "trigonometric functions," the fish-bone diagram whose content value is "trigonometric" is shown. And clicking "exercise" on the fish-bone diagram, an exercise about trigonometric function is displayed. This supporting system is not a type of courseware; a student needs to find what he wants to do and the answer to a selected exercise by himself. In this state, a searching algorithm may allow ambiguous searching.

5 Conclusion

In this work, we attempted to construct a supporting system for the self-learning of mathematics, based on ontology. In order to prepare a resource of this system, we define the Mathtext XML documentation system. Transforming this document, we get Mathtext.RDF. Using this RDF, we intend to construct a supporting system for the self-learning of mathematics. The main principle of this system is that there are sufficient learning attributes for guiding students to study mathematics and a target universe is restricted to one document. This system is able to apply to another subject or field, if its document is structured and has items corresponding to learning attributes of Mathtext documents. There remains an important problem for future research, namely, that of extending *is-a* relations.

Art and Mathematics

A New Pathway

M. Francaviglia, M.G. Lorenzi, and P. Pantano

Mathematics—the universal language of human abstract thought—subtly pervades art, even if sometimes as a rather hidden component. Art and mathematics have, in fact, evolved in a parallel way, along with the changes in our ways of conceiving, perceiving, experimenting, and representing "reality." Paradigmatically, one can divide this evolution into four major steps, from classical art (essentially based on Euclidean geometry) to the modern forms of art (based on the most modern achievements of mathematics). In the era of digital art, a skillful use of web and computer technologies, along with multimedia, offers new challenging ways to introduce mathematics to nonspecialists, starting with a careful reading of artwork. We propose here an innovative teaching project—explicitly aimed at attracting the new generations to the beauty of mathematics—that is mainly based on an innovative web portal called "MArs," in which the deep interrelationship between art and mathematics is used to extract mathematical concepts out of classical and modern pieces of art.

1 Communicating Mathematics in the Twenty-First Century

The tasks of disseminating, communicating, and visualizing the fundamental achievements of modern science are by no means simple ones. Mathematics—the language for excellence of all possible forms of abstract human thought—does not escape from these difficulties. Even if mathematics pervades, in a more or less recognizable way, most of the products of human experience (not only science and technology, for which mathematics is of course the privileged language, but also social sciences and humanities, art itself included), still mathematics is considered by the majority of people as a sort of "Science for Initiates." It is considered a form of human thought that requires special abilities to be understood, essentially reserved to a handful of "gurus" who know its most hidden secrets, that is eventually perceived by the "collective imagination" as a rather tedious discipline. The annoying attitude of certain "modern" mathematical schools to emphasize abstraction rather than the intuitive aspects of mathematical learning has certainly contributed a lot to this widespread misunderstanding. This is true despite the fact that—after what in our opinion has been in the recent past an exceedingly strong abstraction process of mathematical teaching—new didactics and ways to disseminate information are finally arising that tend to revitalize ways of introducing mathematics without renouncing rigor but also without renouncing to mathematics' more aesthetic and directly intuitive aspects (see also [14], and references quoted therein, where the role of gestures in intuitive mathematical teaching and understanding is emphasized). This follows, in a sense, the old but still fresh Latin motto: *"In medio stat virtus."*

Nevertheless, disseminating, communicating, and visualizing mathematics—from its most classical aspects to its newer achievements—is definitely a fascinating challenge. This is not only an interesting and stimulating exercise for the "experts" in the field, but rather an urgent necessity dictated by the new trends of our society. The Italian Ministry of Research, University and Education (M.I.U.R.)—just to mention an experience that is near to us, but nevertheless common to other national and international programs—has also felt the need to help large communities of nonspecialists to approach science and to better penetrate its impact on everyday life. For this purpose, a "Web Portal for Research" has been launched, under the supervision of CINECA (www.ricercaitaliana.it) [247]. It has the explicit aim of reviewing the research initiatives in Italy, and presenting them in fresh ways that are well suited to introduce deep strata of society to the "secrets of research." Among the "specials" that this web portal offers (in Italian), an emblematic one refers to the communication of science. We quote from it [107]:

> Scientists, nowadays more than before, are facing new challenges in the field of Communication. ...Communicating

> Science has nowadays become an absolute necessity. ... Scientists cannot avoiding to interact with a plurality of groups, informing Society and listening to its instances. If on one side they have to better communicate, on the other hand the operators of Communication have to become more critical, more informed about Science together with its mechanisms. ... The European Union considers the ≪Dialog between Science and Society≫ one of the fundamental steps towards the construction of a European space for research. ... The themes related with Communication of Science have therefore to become an essential part in the formation of young researchers ... since there is no deterministic relation between communication activities and the role played by Science in Society and, in any case, there is no Science without Communication. ... Scientists have nothing else to do than discovering and understanding the challenge of complexity in Communication.

And in another part of this "special" one reads [127]:

> A Dwarf. A Dwarf who moves among the tall Titans who possess specialist knowledge. ... A Dwarf able to climb the shoulders of those Giants, to look farther. ... This Dwarf is the "Communicator of Science," an intellectual character who might be still nonexistent. However the new Society of Knowledge needs him more and more to tell herself her own fresh origin and her fast becoming. ... The tall Titans, of course, are mathematicians, natural scientists, historians, philosophers, anthropologists, sociologists who "make" and/or interpret Science, that is the most relevant part of the new knowledge that informs our Society about herself. ... The old walls of the "Turris Eburnea" are definitely falling down and the world of Science finds itself to be completely open towards Society. ... Communicating Science is becoming a more and more diffuse social need. ... We are therefore facing a two-fold necessity. Independent but convergent. Scientists have to communicate with non-experts, to share with them decisions that have a great impact on their own work. ... The world of Science is a complex system with several dimensions. ... The multilayered complexity of the scientific world is interacting with social complexity. ... This mutual relation is modifying our way

> of doing Science as much as the evolutionary dynamics of Society as a whole. ... The system of Mass Communication is evolving and articulating in a such a way not to require and privilege rigid specializations, but rather requiring and privileging flexibility. ... These consequences define the unalienable characters of the Dwarf who is obliged to climb the shoulders of Giants to tell the story of the Society of Knowledge: an intellectual figure who is not lost in the complexity of Science, of Society and of their growing compenetration; someone who, at the same time, is able to ≪govern media≫, i.e. to master the fast evolving and diversifying means of Mass Communication.

Because of the circumstances described above, it seems to us that also mathematics (as well as other more concrete scientific disciplines) is facing, at the turn of the third millennium, the two-fold need of renewing on the one hand its basic teaching—in order to attract more and more people to the beauty and the pervasiveness of mathematics, so to contribute to a much larger diffusion of "basic mathematical thinking" among people—and, on the other hand, creating a "task force" of specialists who not only know the discipline in depth, but are also able to communicate its structure and achievements to a broad audience. As we said above, these specialists must be central figures who, besides having a sufficiently broad and deep knowledge of the main fields of the discipline, as well as of its most promising further lines of development, can help in promoting the dissemination of its specific culture (in our case mathematics) with the aid of well-designed and at the same time rigorously constructed and artistically valid multimedia presentations that can attract an audience toward a deeper immersion in the discipline itself.

It is our strong opinion that a lot can be done in both these directions by relying on the skillful use of all the new tools that the technologies of the Digital Era offer for our ways of teaching and communicating mathematics (and science in general). The potential offered by digital art is in fact well suited to producing objects for scientific communication and visualization in which one can combine high scientific quality together with a rigorous technical frame and a palatable realization enriched by special effects. Thus, there will be a diminishment of all the difficulties that are related to the transmission of an exact scientific message that (as it often happens) might be abstract and rather far from common experience. See [115–117] for a paradigmatic example, where we describe our recent experience with a video and multimedia presentation aimed at visualizing Einstein's Theory of Special Relativity. Our recent experience was stimulated, in fact, by the World Year of Physics (WYP) (see [275]), proclaimed by UNESCO in 2005 to celebrate those papers by Albert Einstein that in the "Annus Mirabilis" 1905 drastically revolutionized the physics and also part of the mathematics of the twentieth century. This emphasized once more that visualizing the discoveries of twentieth-century physics is in fact a fascinating though rather difficult task. It is difficult because the theory is far from standard experience. Relativity has, in fact, provided a deep conceptual revolution that eradicated the classical immutability of "space" and "time" as separate entities, replacing them with "spacetime" (in which space and time are deeply entangled, so that each one loses its own absolute meaning to acquire on the contrary a "relative" meaning that depends on whom is observing and measuring it). Nevertheless, this is also a fascinating challenge, since each one of us would like to tell and explain in simple words the fundamental story of spacetime structure.

We strongly believe that such an experience has taught us much about new ways of undertaking dissemination and communication, as well as approaching new pedagogical pathways based on the use of modern tools. Scientific dissemination can, in fact, profitably use the new media for all the aforementioned purposes, provided the "book paradigm" is suitably reconsidered and replaced by a suitably adapted "laboratory paradigm." Most of the already existing "multimedia books," in fact, contain images, text, and videos, although they are usually structured as "unidirectional tales." In a laboratory, on the contrary, one forecasts and investigates "knowledge," allowing one (or more) mental models to be confronted with reality. Science is born out of this relationship. In a sense, science is this relationship (see also [121] for a short discussion). In our opinion, a dialectic relationship between reality and the investigator is an essential feature of dissemination, that must guide the experience and allow the acquisition of mental models of reality, to be later tested in order to survive or to be rejected. Scientific dissemination is therefore something in between a technical treatise and laboratory research, since notions that are acritically assumed cannot communicate the true essence of scientific experience.

More specifically, talking now of mathematics, we would note what has been recently stated by the Italian Mathematical Union (UMI), in the booklet [2] devoted to indicating a possible pathway for teaching mathematics to "external" users. This booklet was the result of long reflection and consultation with the Italian mathematical community, which ended up in a sort of "decalogue" for teachers at the University level. It deals systematically with the problem of searching for a coherent structure that can be adequately given to courses of mathematics taught at nonscientific faculties and universities, in order to make mathematics more palatable out of its traditional and natural frameworks. There we read in the Preface of [2] (page xvi):

> Mathematics is universally recognized as a powerful language to describe the World, to construct models, to calculate and forecast. Because of this it is considered to be a very useful tool, often indispensable, for a lot of disciplines. ... Also in various disciplines pertaining to Human Sciences, as e.g., the Science of Cultural Heritage and Archeology. ... The teaching and understanding of Mathematics is therefore a relevant cultural and pedagogical problem, that each year concerns thousands of University Teachers and hundreds of thousands of Students. ... To reach the type of knowledge indicated it is not necessary to insist too much on formalism, since it often obscures the meaning of mathematical objects and puts a brake on the development of modeling abilities. The specific competencies ... are better exploited through ... the following abilities:
>
> - reading and interpretation of texts;
> - writing and, more generally, Communication;
> - organization, storage and retrieving of knowledge [...] also by means of informatics devices.

The far-reaching name given to this "mathematical philosophy" is "mattoncini" (i.e., "little bricks"), meaning that any coherent teaching project should consist of sets of small basic pieces to be combined together according to specific needs.

A further experience of ours that points exactly in this direction is a teaching experience that one of us (MF) realized in the academic years 2003–2004 and 2005–2006 at the University of Calabria, in the framework of a PhD course on "Psychology of Programming and Artificial Intelligence" (see the webpage [265] and, more generally, the webpage [264], where ESG activities are reviewed). In exact consonance with the aforementioned aims of "mattoncini," a new course in "Mathematics and Art" has been already tested in the framework of that PhD course; the lectures were attended by a varied audience, including artists. The course was accompanied by a written text (see [108] and [109–112]) that starts with Euclidean geometry and ends up with Riemannian geometry, peeling off the structures that enter our understanding of the Euclidean and sensible world (see later, in Section 3). The starting points are, of course, the definition of geometry as the "art of measurement" and the deductive methods of Euclidean geometry (that can be treated in a critical and adogmatic manner, in line with the logical deductive approach followed by the ancient Greeks, that still to a great extent can be used today to describe, analyze, and understand extended objects in space). Synthetic geometry can be consequently introduced at different levels of abstraction and understanding (set theory, abstract and structured spaces, morphisms and transformations between spaces, topology, Erlangen's program, and Klein's work, in which geometry is defined as the study of the invariant properties with respect to a characterizing group of transformations). Symmetry plays a dominant role in understanding mathematics through art. Other topics that can be conveniently extracted out of art are: curves and surfaces; Euclidean optics (see also [56]); projective spaces; perspective; design; tessellation of surfaces (Picasso and Escher's work; see, e.g., [255]); fractals; knots; crystallographic groups (see [78]); and curvature and non-Euclidean (i.e., Riemannian) mathematics and art (see also [230, 255]).

2 Mathematics, Art and Cultural Industry—A New Pedagogical Challenge

All of what we stated in Section 1 of this chapter refers to a broader field of interest, that can be collectively called the context of "applications of mathematics to cultural industry." By this we mean the whole set of applications of mathematics to "nontraditional" frameworks, i.e., to frameworks that do not belong to (and even apparently are rather far from) the scientific disciplines, where mathematics is—as we already said above—the "leading language" as well as an "unavoidable pattern of thinking."

A strict definition of "cultural industry" is by no means easy. It probably does not exist! In the book [243]—which refers to the birth of cultural industry in Italy—we read:

> One of the most durable meanings of [cultural industry] is its strong relationship with formation and instruction. ... The expression "Cultural Industry" is used for the first time in

an analytic way by European researchers belonging to the so-called "Frankfurt School." ... While intellectuals were discussing, (de facto) an industry of cultural production had already born and developed since a long time.

As we shall discuss in Section 3, mathematics and the art of painting have developed along parallel tracks (see [111, 197]). A fully analogous pathway is easily recognizable in all other forms of visual art—e.g. in architecture (see [225, 226])—as well as in the evolution of music, from its early understanding as the harmonic structure of cosmos to its most modern conception. Moreover, it is worthwhile to mention that in the present day we see mathematics playing a renewed role in all forms of art (visual, plastic, and musical; see, e.g., [20, 96]), with the explicit aim of applying both analytical and geometrical methods to generate art and music by means of computers and electronic devices (see, e.g., [36] and references quoted therein). As a consequence, mathematics is not only an essential tool for science and technology, but also for humanities, in particular for art. And out of art we may say that mathematics gains one of its main reasons for developing and changing in time. Mathematics contributes to our way of conceiving and shaping the world we live in, while art develops the means to harmonize, describe, represent aesthetically—or even to transcend and transfigure—the world of our sensations and perception.

Mathematics, therefore, intersects transversally all forms of cultural industry. In relatively recent years this has been fully recognized and a number of initiatives have been launched to provide forums for discussion about these deep interrelationships. We note some: a periodical series of conferences organized in Venice by M. Emmer (see [95] and references quoted therein); the MAIC Conferences organized by us [33, 35]; and the Generative Art Conferences in Milano (see [240])—in its more recent edition, in 2005, we presented an installation at the crossing of art and science [122], based on an earlier performance of M. Petry first presented at a symposium on Art, Complexity, and Technology (Torino, 2005; [13]). Mathematics enters also the field of "cultural heritage" (see, e.g., the "Speciale" on Pompei eruption [99] in the web portal [247]); we also note the activities of ESG in the field of virtual archeology (see [264]), to which we contributed a proposal concerning the transition from Greek mathematics to modern science [113]. But these are just a few examples. A widespread literature exists on the subject and may be easily found also by relying on quotations from references [95]– [113] above; we particularly note the recently launched journal *Nexus* [273].

With this in mind, it is our strong opinion that a new way of teaching, communicating, and visualizing mathematics can and should be envisaged; a way that deeply requires the understanding of the profound intertwining between mathematics and art. It should be a way, however, that rejects the dogmatic way of teaching mathematics in a classical and somewhat static manner, occasionally and incidentally recognizing its relationship with art; but rather a way in which art is the central theme out of which the existence of mathematical structures is recognized, first as an essential part of the artistic object and only later understood and developed at a deeper level; a way to extract and understand structures, symmetries, and broken symmetries out of their more or less evident appearance within the structure, symmetry, or apparently broken symmetry of a piece of art (see, e.g., [49, 132]). In order to obtain such a different and completely new perspective on mathematics, intelligent use should be made of all the modern tools that new technologies provide us to represent, communicate and visualize science: digital technologies, digital art, computer graphics, computer vision, virtual and augmented reality, multimedia and web technology, and so on. The explicit aim is to revitalize the understanding of the central role that mathematics plays in everyday life, starting not from far-reaching and astonishing scientific results, but from the emotional and aesthetic side of our consciousness and perception. The goal is not only to contribute to a better spreading of mathematics among nonspecialists, but also to promote broader attention from large portions of students to the beauty of mathematics, and even to help specialists to create new pathways for introducing deep mathematical concepts in an easier, more intuitive, more sensible, and more palatable way.

Mathematics, in fact, is less and less known by the young generations. We face a crisis in registrations in courses in mathematics, while in Italy we even see our nation standing low in world scores. Analogously, tests that measure the degree of mathematical knowledge of engineering students in Italy show a negative trend. One of the major problems—in Italy and abroad—is the low appeal of mathematics to new generations. Against a society that privileges images and motion, the teaching standard of mathematics seems to be still attached to traditional schemes that have a low appeal to students. Because of this, we believe that new tools more broadly based on sound and images, as well as multimedia and new communication tools, might help a lot to reverse this negative trend. Exactly in this sense, art—in its multilayered aspects—can play a central role in helping mathematics to find new pathways. The idea we are advocating can be summarized in a few elementary steps:

1. Select a mathematical topic and select a number of artistic objects that are significantly related to that specific subject.
2. Describe the mathematics that is contained or evoked by those artistic objects, with particular attention to the selected topic.
3. Suggest problem solving of an artistic nature to the student, instead of traditional exercises.
4. Have a specifically oriented book (or more than one) where the student can find various sources of further study. The book [108] is just a first step (see Section 4).

All the teaching activities (as well as tutorials and laboratories) should be supported by adequate multimedia tools. Cooperative working frameworks, both "in loco" and at a distance, will thus be of the kind typical of Renaissance Schools.

3 Mathematics and Art: A Parallel Development

As we have mentioned, it is well known and fairly established that mathematics has developed in a parallel (sometimes even precursor) way, not only with the scientific thought, but also with our way of perceiving, describing, and representing the sensible world by means of art. Our cultural history shows, in fact, the closeness of the links between mathematics—a means for discovering and describing reality—and art, which aims to express or represent reality (see [109–112] and [197]). As a paradigm, we may say that the transition from Euclidean geometry to the geometry of perspective in the Renaissance, to non-Euclidean geometry in the eighteenth and nineteenth centuries, up to the development of geometry of "topological forms" in the twentieth century, can and should be seen as a counterpart to the transition from static paradigms of arts and architecture in the antiquities, to the conception of "beautiful painting"—through exactness and depth reconstruction—of Pier della Francesca, to the evolution of artistic shapes in the eighteenth century (e.g., divisionism, expressionism, impressionism) up to the complete destruction of symmetry in the modern, contemporary, and avant-garde forms of art (cubism, fractal painting and so on; see, e.g., [50, 255]). More precisely, the "rettangolo aureo"—the golden mean as well as the theory of proportions—were at the basis of Greek science and geometry; in the Renaissance the artist was a complete man: painter, sculptor, architect, mathematician, and also scientist (as good examples, we may mention Piero della Francesca, Dürer, Brunelleschi, and Alberti). The idea of an exactly Euclidean world and the need for painters to represent faithfully the three-dimensional world in just two dimensions eventually led to the birth of projective geometry (in a sense, the use of the point at infinity as an ordinary point). The discussion about the validity of Euclid's fifth postulate led, however, also to hyperbolic geometry and, in the artistic field, that paved the way for the development of impressionism. It turns out, in fact, that our visual space is hyperbolic rather than Euclidean (Lunemberg, 1947; see [197]), so that it is not surprising that artists in the eighteenth century began to represent what the eye actually sees rather than what the eye is presumed to see in a fully Euclidean world. The nineteenth and twentieth centuries have finally seen the introduction of time as a fourth "sensible" dimension alongside—and on equal footing—height, length, and depth (Einstein's Theory of General Relativity is at the top of this line of thought); motion and curvature become part of the world and not something which is embedded into the world. This new perspective supplies objects with a dynamics impossible to represent in just two or three dimensions; yet artists (e.g., Balla, Boccioni, Duchamp) tried to provide pictorial representations of movement. Dynamism in modern art is finally achieved through the moving camera and cinema. The idea of letting sections in one dimension flow less in time in order to represent objects in one dimension more may be also used in figurative arts (think of Picasso or Dalì, and a four-dimensional hypercube that opens up in three-dimensional space). The new mathematics of the twentieth century is also the mathematics of manifoldness, curvature, discreteness, fractals and chaos—and some modern ways of painting reflect these new ideas (Picasso, Pollock, and so on). Also, the famous artist Escher, the unchallenged inventor of impossible objects and imaginary worlds, was influenced by the mathematical theories of Poincarè and Penrose in creating his striking pieces of art. The connection existing between art and mathematics is therefore universally recognized and needs no more examples. Several papers on the above subjects can be found in the collection [95, 96, 273].

4 The Project "MArs"

The PhD course and the book [108] that we mentioned at the end of Section 1 (based on the historical parallel described in Section 3) may be precious tools for deepening the teaching, and the understanding, of mathematics along the lines indicated above. However, in our opinion, a much broader project has to be developed—a project that changes completely the viewpoint and uses art as a first theme to develop a modern communication, visualization, dissemination and teaching of mathematics, at all possible levels and for all possible applications. We are therefore projecting and developing an innovative "full teaching project," mainly (but not only) associated with a web portal that is dedicated to "Art and Mathematics." This is a new project aimed at allowing the progressive understanding, at progressive levels of deeper and deeper mathematical reasoning and abstraction, in which art comes first and touches the emotions, thus stimulating the need and the desire to penetrate more intimately into the structures that underlay art itself, often without revealing themselves in an explicit way.

The web portal [118, 267] which is a central object in this project and, as such, the first step toward a full development of the entire project, will therefore present art as a way to approach mathematics, to enjoy the beauty of the structures existing in our vision and representation of the world, and eventually to reconstruct the structures and theoretical tools that are necessary to understand and elaborate their true essence, as well as to show or stimulate interdisciplinary further applications to other fields in which mathematics plays a key role. The portal will propose artistic objects and use them to discover their (usually hidden but sometimes evident) mathematical richness. It will propose a reasoned selection of links to other (usually differently oriented but extremely important) websites (as, e.g., the website [266]), and also to information about initiatives in the field of art and mathematics (conferences, exhibitions, courses, and so on). The portal will also exhibit a collection and a review of multimedia explicitly produced for the visualization of mathematics (and, more generally, science), as well as display materials on the appropriate use of web technologies aimed at implementing the interactions between mathematics and art. In the future it will also contain an electronic newsletter (in which both innovative and more traditional ways of communicating mathematics through art will find an appropriate forum). We think, at a first stage, of visual and digital art, music, architecture, biology, visualization, e-Learning; but other fields are worthy of being explored, too. Such a newsletter may also contain "nontechnical" papers, or extended reviews, about collateral aspects, giving space as contributors both to pure mathematicians and to specialists of other fields interested in the interaction between mathematics and art. The portal will also contain a less conventional space dedicated to "work in progress," such as: presentation of multimedia and/or innovative ideas; simulations and generative approaches to "mathematical art"; digital technologies to produce art through mathematics or to understand mathematics through art; and new frontiers in mathematics stimulated by digital art and artificial life, virtual, and augmented Reality. The portal will finally be able—after a first period of grounding—to promote a service of expertise and counseling for mathematicians and nonmathematicians about ways to better implement their presentations for lessons, lectures and conferences that require an interplay between art, visualization and mathematics.

The project is being developed with the help of a scientific committee, currently under formation, representative of both mathematics and other fields (not only in art), so to be able to: 1) broadly cover a wide range of interests; 2) open up a real dialectic between mathematics and humanities (art, architecture, music, and so on); 3) stimulate interdisciplinary attention to new and innovative products that exults the traditional competence of single frameworks and, as such, cannot be easily classified under the traditional disciplinary classifications; 4) and give adequate space to new tendencies and new problems generated by web technology.

The explicit aims are: 1) to give preferential attention to the most stimulating, innovative, and interdisciplinary products (larger space not to demonstrations or to self-referential artistic performances, but rather to critical discussions about the infinitely many relations between art, architecture, mathematics and humanities); 2) to create an interactive dialog between different "poles," a dialog that is becoming of strict actuality and is made possible exactly by web technology and digital art; 3) to strongly repropose new ways of interaction between art, science, and technology; the key role of "open minded" mathematicians will be crucial, along with the role of "mathematically inclined" artists, since only out of a fruitful interaction between these two kinds of communities—that the portal will help to stimulate—new and innovative approaches to communication, visualization, teaching and computer graphics will be eventually able to emerge; 4) to try to eventually recreate the fruitful "constructive and internal interaction" that existed between mathematics and art in the Renaissance or in the Dutch painting of the seventeenth century; a liaison that, in a sense, has been interrupted and replaced by "external" interactions between the two disciplines, in which a dialog existed in more recent times, in our opinion, only at a marginal, secondary, and less promising level of efficiency. The future will tell us if we are right.

A List of C2C Past Talks

The presentations to date in the Coast-to-Coast seminar have been a mix of mathematical and computational talks, with a wide variety of focuses within each field. As of fall of 2006, the C2C seminar has featured the following presentations[1]:

- Sherry Mantyka (Director, Mathematics Learning Centre, Memorial University), *The Math Plague: Learning Strategies for Under-Achievers in Mathematics* - December 2006, presented from Memorial University
- Gordon E. Swaters (University of Alberta), *Modelling Deep Ocean Currents* - November 2006, presented from the University of Alberta
- Ken Barker (University of Alberta), *Privacy Protection in Large Data Repositories* - November 2006, presented from the University of Alberta
- Laurence T. Yang (St Francis Xavier University), *Scalable integer Factorization for Public Key Cryptosystems* - November 2006, presented from ST. F.X. University
- Jonathan Borwein (Canada Research Chair, Dalhousie University), *Notes from the Digital Trenches* - October 2006, presented from D-Drive
- Steve Thompson (Shrum Chair in Science, Professor in Statistics, Simon Fraser University), *Sampling in Networks* - September 2006, presented from IRMACS
- Ron Fitzgerald (President, Math Resources Inc), *Learning Infrastructures and Content Authoring* - March 2006, presented from the Halifax offices of Math Resources Inc

[1]This includes the first full academic year of the Coast-to-Coast seminar, with the second year currently in progress.

- Bojan Mohar (Mathematics, Simon Fraser University), *Hadwiger's Conjecture* - March 2006, presented from IRMACS
- Carey Williamson (Computing Science, University of Calgary), *Things that Go Bump on the Net* - March 2006, presented from the University of Calgary
- Jeff Hooper (Mathematics and Statistics, Acadia University), *L-Functions and Arithmetic* - February 2006, presented from Acadia University
- Alejandro Adem (Mathematics, University of British Columbia), *Periodic Complexes and Group Actions* - February 2006, presented from IRMACS
- Przemyslaw Prusinkiewicz (Computer Science, University of Calgary), *Computational Biology of Plants* - February 2006, presented from the University of Calgary
- János Pintér (Pinter Consulting Services), *Teaching OR/MS Using Integrated Computing Systems* - January 2006, presented from D-Drive
- Jonathan Schaeffer (Computing Science, University of Alberta), *Solving Checkers* - January 2006, presented from the University of Alberta
- Andrew Rau-Chaplin (Computing Science, Dalhousie University), *Parallel Applications in Phylogeny* - December 2005, presented from D-Drive
- Arvind Gupta (Computing Science, Simon Fraser University), *The Inverse Protein Folding Problem* - November 2005, presented from IRMACS
- Karl Dilcher (Mathematics and Statistics, Dalhousie University), *Wieferich Primes and Fermat Numbers: Computations and Generalizations* - November 2005, presented from D-Drive
- Uwe Glaesser (Computing Science, Simon Fraser University), *Semantic Blueprints of Discrete Dynamic Systems* - October 2005, presented from IRMACS
- John McHugh (Computer Science, Dalhousie University), *Pyrite or Gold? It takes more than a pick or shovel* - October 2005, presented from D-Drive
- Peter Borwein (Executive Director, IRMACS, Simon Fraser University), *The Riemann Hypothesis* - September 2005, presented from IRMACS
- Jonathan Borwein (D-Drive Director, Computer Science, Dalhousie University), *Mathematical Visualization and Other Learning Tools* - September 2005, presented from D-Drive

B Guidelines for Managing a Distributed Seminar

Organizational Issues

The growth of the C2C Seminar significantly increased the number of people involved in the organization and the running of the seminar. To guarantee full technical support, each site needs a technician present locally during every presentation. Since WestGrid and ACEnet employ most of those people, the seminar heavily depends on the willingness of the two institutions to be involved in the C2C project. In addition, for recruiting speakers and advertising presentations locally, it is important to have C2C liaisons at as many universities as possible. Clearly, the liaisons should be faculty members ready to put their valuable time into the process of creating something that is as new and complex as the C2C series. However, as the size of the core group increases as more universities become involved, the overall organization becomes more complex. Thus, ideally the number of people involved in organization of the C2C Seminar at all levels should be about two dozen. The size of the group immediately raises the question of the hierarchy and communication within it. Currently (November 2006), we are working on the establishment of a cross-institutional email list that would serve as a forum for all people involved in running the series. Also, in an effort to centralize the process of coordinating all of the people involved, the position of Seminar Coordinator will be introduced starting in the Spring 2007 term. The main goal is to centralize scheduling and announcement distribution, and to simplify communication between the C2C group and the rest of the academic community.

One of the challenges in coordinating a series of distributed events across a large geography like Canada is the need to constantly consider time zones. There are six different time zones from British Columbia on the West Coast to Newfoundland in the East. This restricts scheduling possibilities, for a seminar scheduled at the reasonable hour of 11:00 am in D-Drive on the East Coast would be a little too early for audiences on the West Coast at 7:00 am. Seminars are presented therefore at 3:30 pm Atlantic Time, which is 11:30 am Pacific Time, and 4:00 pm Newfoundland Time.

The issue of time zones also affects communications and advertising between West and East: there are ample opportunities for misunderstanding unless each stated time is labeled explicitly with the local time zone. To state times in terms of your own local time zone and rely on the recipient of communications to translate into their own local time has worked the best for us. Early attempts to translate communications into every applicable time zone resulted in an increase in errors and misunderstandings.

Another issue that has arisen from having multiple academic institutions involved in this event is the fact that scheduling around the academic timetable becomes potentially more complex. It is difficult to guarantee that a given seminar will not conflict with other activities at one or more universities. We also need to have the local WestGrid and ACEnet nodes be available at the time of our talks. Our solution to these potential difficulties is simply to choose a regular time at the outset and stick to it, with the understanding that there may be conflicts with the local academic schedule, but that the regularity of the event gives some possibility for math departments, computer science departments, or other regular attendees to organize their schedule to accommodate the Seminar Series. We are also able to record the entire event and to allow it to be watched by a class or other interested group at another time.

Finally, advertising the seminars is complicated by their distributed nature. It is necessary to make clear on any particular advertisement both the time and the place of an event; however, the time depends upon the local time zone, and there are sometimes several places at which the presentation can be attended locally. Our solution is to distribute communications through a hierarchical system: once the details of an event are determined and agreed between the IRMACS and D-Drive administrators, then each of them distributes the information to their respective coasts. In D-Drive, this means that the administrator is responsible for passing the information on to St. Francis Xavier University, Acadia University, Memorial University, and other interested parties. At each of these final destinations the recipients advertise to their local audience using local time and the local point of attendance. A central listing of Seminars is also maintained on the D-Drive website: www.cs.dal.ca/ddrive/seminars. Similarly, on the West Coast, communications move outwards from the IRMACS administrator to the other universities and each advertises locally. The announcements are also posted on the WestGrid website together with the list of all seminar locations in Alberta and British Columbia. As there are currently only two time zones and three regular points of attendance on the East Coast, a joint poster has been designed by ACEnet's Chief Technical Officer, Greg Lukeman, at St. Francis Xavier University, that carries the multiple time zone information and multiple attendance locations. It has the advantage of presenting a consistent image to the whole Eastern community at the expense of being a slightly denser packet of information. A version of the poster for Western Canada is made and distributed by WestGrid.

Organizing Content

The Coast-to-Coast Seminar Series has arisen from a partnership between the two founding members, IRMACS and D-Drive, and although it has grown to include other participants, it still retains that basic organizational hierarchy. Planning decisions are made bilaterally between the two founders and then passed on to the other members. At the beginning of the academic year a timetable is drawn up for the series, with presentations alternating between East and West. For the Eastern series of presentations, the Director of D-Drive, Dr. Jonathan Borwein, then compiles a set of suggested speakers, who are approached by the administrator until a full roster of speakers has been confirmed. The director will also indicate a desired distribution of speakers from the other Eastern universities and then the task of finding a speaker at a remote site passes to a representative of that site.

A protocol for organizing the process between D-Drive and IRMACS was drawn up at the beginning of 2005 and has served as a useful guideline, although it has not always been possible to follow it to the letter. It includes the following procedures.

1. Two–three weeks before the lecture:
    - Confirm the booking location.

- Get the title and abstract of the talk, send them to all remote sites, and post them on the web.
- Inform the local technician about the date and time of the lecture.

2. One–two weeks before the lecture:
   - Contact the presenter and find out how he or she will deliver the lecture; acquire a copy of the presenter's material[1] to be used in testing; forward the presentation to the local technician.
   - Check if the technician has confirmed testing times with all remote sites.
   - Send out the first email notice to all sites to invite people to the lecture.
3. Two days before the lecture:
   - Final lecture reminder to listserves and contacts on campus and surrounding institutions.
   - Final consultation with the technician.
   - Reminder email to all sites that we start set up one hour in advance.
   - Final email to presenter for any last minute concerns.

In practice, much of the process has become automatic. Sometimes, however, it proves difficult to pin down a speaker far enough in advance and the whole process compresses into a shorter time frame. Some speakers have to be reminded quite vigorously to submit a title and abstract well in advance.

In D-Drive, and in IRMACS, the administrative role and the technical role are separate. This has also led to some adjustment in the protocol, as it often makes more sense for the technician to communicate directly with a speaker about testing issues or about the speaker's presentation materials if there is a technical question to be addressed. Good lines of communication between the administrator and the technician are important for good co-ordination of these activities, and in D-Drive these two people inhabit the same general lab space to facilitate communication.

[1] While a lecturer in D-DRIVE could, in principle, write directly on a SMART Board, we discourage this (except, say, in answering questions or for annotation), as it takes a good deal of experience to do this effectively. Likewise, we do not encourage writing directly on a projector or something similar. Advance preparation is really advisable, and digital transmission is preferable to analog.

General technical issues with distance collaboration. Technical challenges discovered while providing a distance collaboration experience such as the Coast-to-Coast Seminar are often strongly related to policy, organizational, and educational issues in the various institutions. The goal of the C2C Seminar from a technical perspective was to build a collaborative seminar experience in which the technology is a transparent mechanism for communication, rather than a barrier or limitation to communication. With that goal in mind, familiarity with the available technology at each of the sites becomes a critical component to building policies and procedures that result in a technologically seamless seminar experience.

Site preparation and testing. As an introduction to general technological issues, we cannot emphasize strongly enough the importance of testing and site preparation for the seminar experience. Each institution should have a local technician on site who is both familiar with the site equipment and has conducted thorough tests with other participating sites to test their equipment. To facilitate site preparation, checklists for site participation should be developed, and reviewed regularly by the on-site technician for an institution. The heterogeneous nature of site equipment will result in some variance of individual site checklists, but all site checklists should generally cover the following areas:

- stability of hardware used for the site,
- reliable client connection to the venue server,
- audio quality tests, both receiving and sending,
- video quality tests, both receiving and sending,
- desktop sharing tests, both receiving and sending.

Institutions presenting a lecture have an additional set of tests that should be run in advance of the presentation date. The first is that the presentation content should be initially reviewed locally on the presenting equipment to ensure that there are no issues with display or audio anomalies. This gives speakers an opportunity to amend their presentation, or for local technical staff to make adjustments to the presenting system.

In addition, the presenting site should review with remote sites the quality and speed of the presentation materials running over the desktop sharing software. Current desktop sharing software can suffer from update delays that may be acceptable for a static slideshow, but that cause the software to be completely unusable for a video clip inserted into a presentation, a highly graphical presentation, or a local program animation. Advance review of the presentation materials with remote sites will catch technology limitations with desktop sharing software and allow for one of the following:

- distribution of the material to all sites for local execution of the material;
- use of an alternate remote distribution technology, such as the Video Lan Client (VLC), to broadcast presentation videos;
- adjustment of the presentation to account for the technology limitation.

Finally, the presenting site should ensure that speakers have been given an introduction to the presenting system and the technology tools available to them. The interface for items such as whiteboard drawings and presentation controls, and how they interact with remote sites, should be reviewed with speakers before the presentation date. Speakers should also be made aware of how to highlight their presentation in such a way that remote sections of their audience can see it. This has been demonstrated vividly to seminar audiences on several occasions in which the speaker pulled out a laser pointer partway through the presentation and proceeded to highlight sections of the presentation while lecturing, completely forgetting that the remote audiences would not be able to see the indicator. In such situations, it is a delicate decision as to when it is more disruptive to the seminar to correct the speaker, or to live with the problem. Such decisions are often debated and made by the technical group using an instant messaging client in the background during the presentation.

Live technical monitoring. After thorough site preparation and testing, the next critical factor for technical success is live technical monitoring at each site during the presentation. Adjustments will periodically need to be made to technical equipment throughout a presentation, and in a distributed lecture, we do not have the option of just using the blackboard for the remainder of the talk. A technician should be available at each site to provide feedback to the other sites on technical issues and to intervene locally if there is a significant issue.

Technical Compatibility Issues Between Sites

Network compatibility. The most common issues seen when first getting a site set up for the Coast-to-Coast Seminar are with the site's network. Access Grid uses a network protocol called IP multicast to communicate audio and video information to all sites. This protocol allows for a much more efficient communication of audio and video information from a site, greatly reducing the network bandwidth required to send your audio and video information out to multiple sites. If a site's network is not capable of supporting IP multicast, that site will not see video or hear audio, even if it is successfully connected to the AG venue server.

There are two common solutions for this problem: talk to the institutional network engineers about enabling IP multicast on that site's network, or connect to the other sites using an intermediary site (or proxy) that does support IP multicast, called a "unicast bridge." The Access Grid client has a built-in utility to connect to unicast bridges for a venue server, and we recommend starting troubleshooting with that utility when a given site cannot hear audio and see video from other sites. Lack of IP multicast support on the network is the most common problem when first setting up a site, and oftentimes IP multicast will not be an option at all when connecting through a commercial Internet service provider.

The other common connectivity issue for new sites is that a firewall is set to block the network ports used to communicate audio and video information. The ports used to transmit audio and video information are determined by the Access Grid venue server. Those ports will need to be "opened" or "unblocked" for audio and video communication. In the case of the Coast-to-Coast Seminar, we have a document created by WestGrid (our venue server hosts) to send to new sites that lists the port ranges that need to be opened for Access Grid collabora tions.

While not common yet, one network issue that will create additional complexities in the near future is the expanding use of NAT (Network Address Translation) devices. These devices are used on networks to allow many different IP addresses to share a single IP address with an outside network. While these are commonly used now on small home networks (often as part of a "broadband router"), there is an expanding role for such devices in many IT organizations. NAT devices require some complex "port forwarding" scenarios for Access Grid clients, and we recommend avoiding such devices on networks that will have Access Grid clients.

The final network issue that should be considered is available bandwidth. The average Access Grid session will generate approximately 800 Kbps per camera being broadcast. For general planning purposes, we recommend having approximately 15 Mbps of bandwidth available for collaborative seminars such as the Coast-to-Coast Seminar. This is usually not an issue on most university networks, but it can become an issue when sites are connecting over a DSL line or other home broadband connections. In our experience, cable modem services of 5 to 10 Mbps can support collaborative sessions fairly well, but slower connections (such as the average DSL connection) will have issues with bandwidth.

Client compatibility. Several technical compatibility issues have been discovered while attempting to integrate various Access Grid and InSORS clients into the Coast-to-Coast Seminar since its inception. Currently, the Coast-to-Coast Seminars are standardized on the Access Grid 2.4 client, and the recently released Access Grid 3.0 client is undergoing testing at some sites. Our recommendation for collaborative seminars in general, based on the Coast-to-Coast experience, is simply to watch which versions of video-conferencing software a site is using, and have all sites standardized on the same version of the client software.

When using multiple client types for a seminar (such as Access Grid and InSORS clients), it is critical to ensure that your audio and video streams are using codecs that all sites can understand. This often means that the highest common standard is selected for audio and video. In the specific case of the Coast-to-Coast Seminar, we have standardized on the H.261 video codec for all sites. A common issue for sites in our seminar series using the InSORS Windows client is that InSORS video has been set by default to a higher quality codec (H.264) that cannot be decoded by Access Grid clients, or InSORS clients running on MacOS.

The final issue to address in the area of client compatibility is desktop sharing software. We have already mentioned that the Coast-to-Coast Seminar organizers selected VNC software due to its cross-platform availability. Even with a platform-independent software utility such as VNC, there are compatibility issues to watch out for. VNC clients are not universally consistent in their features across platforms. For this reason, we have developed standards to address desktop sharing compatibility.

One example of desktop sharing standards is simply to limit the presentation screen resolutions to standard sizes. Some sites have high definition television screens for their presentation systems, while others have standard portable LCD projectors. If a presenting site sends out presentation information in a high-resolution widescreen format, it becomes highly problematic for sites with smaller resolution displays. While some VNC clients can simply scale the display to an appropriate video resolution, VNC clients on other platforms do not have that capability. For that reason, the Coast-to-Coast Seminar has standardized all main presentation displays at 1024x768, and limits applications such as whiteboard areas to standard projector sizes.

Audio/Video Issues

Audio production. Over the course of the first year of the Coast-to-Coast Seminar, audio clarity was identified as the most critical single component for a remote collaboration experience. Video and desktop presentation quality can fluctuate and even have intermittent interruptions without a significant loss to the quality of a lecture, but audio quality loss almost immediately affects the effectiveness of the communication.[2]

For that reason, we strongly recommend that some careful thought be given to both audio components and room design in a larger Access Grid site. A quiet environment in which outside noises are minimized is critical for any large speaking environment, but in a large video conferencing environment additional consideration must be given to limiting audio noise within the environment. This is simply because the echo-canceling technologies available today for video conferencing have limitations in their efficiency, and the technology usually involves frequency cancellation and noise reduction algorithms. If too many frequencies are generated by a site due to background noise, echo cancellation units can end up canceling too many frequencies and interfering with communication.

To minimize this background noise effect, we recommend that boundary microphones only be used for Access Grid sites with fewer than eight people, and that directional handheld or lapel microphones be used for sites with more individuals. We have also established protocols for the Coast-to-Coast Seminar to minimize background noise during presentations—for instance, all remote site microphones must be muted for presentations other than during specific question and answer sessions.

Controlling audio quality is the primary task for technicians monitoring live Coast-to-Coast Seminars. Oftentimes the lecturer's volume

[2]Our experience certainly is consistent with hearing-disabled advocacy groups' frequent assertions that deafness is more isolating than blindness in modern society.

or the audio environment will change throughout the presentation, and local technicians are responsible for adjusting audio levels to compensate. In this task, they are serving the same function as an audio technician for any other production, whether music, theater, or speaking—but they have the additional complexity of monitoring issues created by the distributed nature of the communication. In these situations the feedback of remote site technicians becomes crucial, as there are times that audio distortion can occur remotely (due to microphone levels overdriving slightly, combined with packet loss on the network) when it would not be noticed by a local technician.

Video production. The Coast-to-Coast Seminar series has also highlighted the usefulness of video production methods as an area of expertise for technicians involved in remote collaboration. Minor production adjustments in the video presentation can make a vast difference in the perceived quality and effectiveness of the seminar experience. These video production adjustments can be items as simple as:

- *Camera location.* Place cameras in such a way that speaker movement is captured from optimal angles and distance, and in locations where it appears the speaker is looking into the camera when facing his presentation materials and remote audiences.
- *Background and lighting issues.* A common problem when setting up sites is an incorrect lighting environment for the cameras to pick up a speaker well. Here is one example from the D-DRIVE Lab that illustrates lighting issues well: when lecturers in the D-DRIVE Lab would stand at a SMART Board video monitor to write on one of the whiteboards, the brightness of the white video display behind them would cause the cameras to wash out the speakers completely. By simply changing the onscreen "whiteboard" to a "blackboard" (changing the background color of the application from white to black), the video cameras were able to adjust their brightness settings and send out a viewable image of the lecturer writing at the board.[3]
- *Remote speaker location.* Display of the speaker and their presentation should be oriented so that the speaker, when indicating their local presentation content, is also gesturing in the direction of the content at remote sites. Remote audiences find it highly distracting if the speaker and content windows are placed on displays at remote sites so that the speaker is gesturing in the

[3]This did also necessitate providing white virtual chalk.

"wrong direction" at the content. The degree of discomfort experienced by remote audiences in these cases has been surprisingly strong.

- *Remote audience location.* Displays of the remote audience should be placed in an optimal way for the speaker to see activity and questions at remote sites while still seeing his local audience and presentation materials. This area of video production also affects room design.

Technical Communication

Communication between site technicians is organized into several established protocols for the Coast-to-Coast Seminar series. Initially, the lead technician at the site hosting the lecture will send a communication to all site technicians giving the following information:

- venue server and room location with instructions for each client;
- presentation details and any special instructions on lecture materials;
- information on connection testing times and the availability of the hosting site;
- "back channel" communication methods for technicians when monitoring the live seminar (usually via a chat server such as Jabber).

All sites attempt to schedule a test run during a regular meeting time the day before a C2C Seminar, to ensure that all site audio/video systems are working correctly, and that any presentation issues are identified and addressed beforehand.

During the lecture, technicians at remote sites initially give feedback to the hosting site technician about speaker audio levels, presentation material transition times, video framing, and any other feedback that the hosting site technician can use to make adjustments that improve the quality of the seminar experience. Technical staff also communicate any production issues to all other site technicians as they occur (such as a disconnect from the VNC desktop sharing software), so that other technicians know to watch for a potential problem.

Emergency Planning

Technicians involved in the Coast-to-Coast Seminar series are also encouraged to have specific plans to deal with common emergency scenarios. For instance, a good plan for sites hosting lectures is to have a procedure for dealing with battery failure on the speaker's mike in an unobtrusive way. Sites are also encouraged to be practiced in interventions such as blocking audio feedback loops by immediately muting all sites other than the presenting site. For "worst case scenario" situations, sites are encouraged to have a backup communication method for both the technicians and the audience. This can be as simple as having phone contact information for the hosting site and a speaker phone available.

For instance, in the event of unknown audio interference during the speaker's lecture, the affected site(s) would immediately mute all audio other than the speaker's, and send a note to other site technicians describing the problem. Other sites would check their audio output to make sure that they were not inadvertently broadcasting the interference. If one site was identified as the cause of the interference, the other sites would block all audio from the source of the problem until that site's technician assured them that it was safe to resume normal communications.

An additional communication area for the hosting site's technician to consider is how to communicate with the speaker if there are technical issues with the presentation/equipment, or the speaker forgets how to navigate a portion of the interface midway through the lecture. As with other emergency scenarios, communication with the speaker in emergency situations should be focused on minimizing or resolving the technical issue as quickly as possible, and eliminating the distraction from the lecture for the audience.

Lessons Learned

Here are the technical consideration and recommendations for managing a distributed seminar:

- Audio clarity is the most critical component of a successful seminar series, and should be given top priority when selecting audio equipment and designing room layout.
- Site technicians for larger sites should have a technical background in audio technologies and sound systems as well as computers and networking.
- Site technicians should always have full access to the audio system and software control panels for audio throughout the seminar.

This should be considered when designing the site and setting up client services on the site PCs.

- Hosting sites should check audio levels in both the audio control panel and via feedback from remote sites shortly after a lecture starts. It is very common for lecturers to increase in volume from initial "mike-check" levels when they start actually presenting. This can cause the audio system to overdrive and introduce distortions in the audio at remote sites.
- A directional microphone is strongly encouraged for lecturers (whether lapel or handheld), rather than having lecturers use boundary microphones. The audio quality of the lecture will be reduced when using boundary microphones because lecturers often turn their heads away from the microphone location.
- Lapel microphones for lecturers should be clipped centered on the presenter's shirt, not clipped to one jacket edge. If the microphone is on a jacket edge it will have a tendency to create audio fading issues as the speaker changes the direction he is facing, and it is quite common for the jacket to fall in such a way that it covers or muffles the microphone.
- All remote audience sites should have microphones muted during presentations until designated question and answer periods. This will greatly minimize site interruptions and distractions in a distributed seminar, as the distracting sounds picked up by microphones will come from the same speaker system as the lecturer's voice (rather than from behind or to the side). This will make it harder for audience members to filter the distracting sounds, and therefore should be minimized.
- Any sound errors introduced by a site should be immediately isolated (site muted) until they can be resolved behind the scenes or after the talk. Remote technicians have the capability to mute any individual site from their audio control panel, and this should be their first reaction to an unknown audio issue.
- When using Access Grid, the Linux client tends to be more stable for audio services than Windows clients.
- Windows-based video capture devices are often limited to one camera per PC system, due to drivers. Care should be used when selecting video capture devices for Windows systems if you need more than one video camera per system.
- H.261 video is a standard codec to allow video compatibility across platforms (Windows, Linux, Mac) and clients (Access Grid/InSORS).

- DV (digital video) recorders with Firewire (1394) output are recommended for sites purchasing new cameras. This will allow those sites to start experimenting with high-bandwidth/high quality video streams when using extensions to Access Grid, such as Ultra Grid.
- When using VNC for presentation, it is recommended that the VNC server be set to disallow remote mouse and keyboard control. This will keep a remote site from inadvertently stealing the focus from the lecturer when adjusting a remote display.
- If using a VNC reflector, all VNC clients must be set to "shared." Clients not set to "shared" will experience disconnects immediately after a connection occurs.
- To ensure that all audiences can see pointer events during a presentation, presenting sites are encouraged to purchase a remote pointer mouse and instruct lecturers in its use and the necessity of using onscreen pointers during a distributed lecture.

C Curriculum Guideline and Mathtext Example

Part of Curriculum Guideline

In junior high school mathematics

Triangle

Right triangle—Pythagorean theorem

Similar triangle—Theorem of similarity

In high school mathematics

Mathematics I (first year)

Triangle and trigonometry

Definition, Basic formula, Relations with geometric figure

Mathematics II (second year)

Various functions

Trigonometric function, Angle extension and radian measure, Properties (graph, periodicity), Angle additive law Double angle formula, Half angle formula, Transform between sum, difference, product

Mathematics III (third year)

Analysis

Limit (Trigonometric function—radian), Differentiation, Integration

Mathematics B (second year)

Complex number

Polar form, de Moivre formula, n-th root

Mathematics C (third year)

Matrix algebra—Rotation matrices

Curves and surfaces

Parametric curve—polar coordinate

An Example of Mathtext Document

```
<?xml version="1.0"?>
<!DOCTYPE mathtext PUBLIC "some url">
<mathtext><text_info> author, etc. </text_info>
<subject_info category="MathematicsII"/>
<chapter content="trigonometry (advanced)" background="trigonometry (elementary)">
  <title>Trigonometry</title>
  <section content="trigonometric function">
  <preparatory value="trigonometry (elementary)"/>
    <title> Trigonometric Function </title>
    <subsection content="general angle">
       <title> General Angle </title>
    <preparatory value="acute angle"/>
    <preparatory value="obtuse angle"/>
       <title> General Angle </title>
       <division att="definition">
          definition of general angle over 180 degree and under 0 degree
       </division>
       <division att="example"> sample exercise </division>
       <division att="exercise"> small exercise </division>
       <division att="definition">
          <title> Radian Measure</title>
             definition of radian
       </division>
       <division att="exercise">
          <numberedList>
             <listItem> exercise 1 </listItem>
             <listItem> exercise 2 </listItem>
          </numberedList>
       </division>
       <division att="proposition">
         <title> Arc Length and Fan Area </title>
       </division>
       <division att="exercise"> small exercise </division>
   </subsection>
   <subsection content="trigonometric function(radian measure)">
    <title> Trigonometric Function </title>
    <preparatory value="generalAngle"/>
    <division att="demonstration">
      definition of sine, cosine and tangent for general angle
    </division>
    <division att="example">
       examples of sine, cosine and tangent
     </division>
    <division att="exercise"> small exercise </division>
    <division att="example">
       examples of sine, cosine and tangent
    </division>
    <division att="definition">
       definition of quadrant
    </division>
    <division att="exercise"> small exercise </division>
  </subsection>
  <subsection content="formula(basic)">
    <title> Property of Trigonometric Function </title>
       Basic relations between sine, cosine and tangent
    <preparatory value="radianMeasure"/>
    <division att="proposition">
       statement of proposition
    </division>
```

```
    <division att="exercise" class="sample">
       sample exercise
    </division>
    <division att="solution"> solution of exercise
    </division>
    <division att="exercise"> exercise </division>
       Periodicity
   </subsection>
   <subsection content="trigonometry (graph)">
     <title> Graph of Trigonometric Function </title>
  </subsection>
  <subsection content="equation and inequality">
    <title> Equation and Inequality </title>
    <division att="example"> example exercise </division>
    <division att="exercise"> exercise
      <numberedList>
         <listItem></listItem><listItem></listItem>
  </numberedList></division></subsection>
</section>
<section content="angle additive law" background="trigonometry (advanced)">
   <title> Angle Additive Law </title>
   <subsection content="angle additive law">
      <title> Angle Additive Law</title>
      <division att="proposition"> additive law
       </division>
      <division att="proof"> proof of additive law </division>
   <division attt="example"> example </division>
   </subsection>
   <subsection content="double angle formula">
      <preparatory value="angle additive law"/>
      double angle formula, half--angle formula, transform
      between sum, difference and product
   </subsection>
</section>
<section content="exercise" background="trigonometry">
   <subsection content="exercise">
      <title> Exercise </title>
      <division att="exercise">
         <numberedList>
            <listItem></listItem><listItem></listItem>
         </numberedList></division>
   </subsection>
</section>
</chapter>
</mathtext>
```

Bibliography

[1] A Maths Dictionary for Kids. Available at [url:169].

[2] G. Accascina, G. Anichini, G. Anzellotti, F. Rosso, V. Villani, and R. Zan. "La matematica per le altre discipline—prerequisiti e sviluppi universitari." *Notiziario U.M.I.* 1(Anno XXXIII):114 (2006).

[3] Access grid community site. Available at [url:65].

[4] T. Ager, C. Johnson, and J. Kiernan. "Policy-based management and sharing of sensitive information among government agencies." In *Proceedings of the 25th Military Communications Conference (October 23–25, Washington DC, USA)*, IEEE, 2006.

[5] R. Agrawal, R. Bayardo, C. Faloutsos, J. Kiernan, R. Rantzau, and R. Srikant. Auditing compliance with a hippocratic database. In *Proc. of the 30th Int'l Conf. on Very Large Databases (Toronto, Canada, August 31–September 3)*, Morgan Kaufmann, 2004.

[6] R. Agrawal, J. Kiernan, R. Srikant, and Y. Xu. "Hippocratic databases." In *Proc. of the 28th Int'l Conf. on Very Large Databases (Hong Kong, China, August 20–23)*, Morgan Kaufmann, 2002.

[7] F.K. Aldrich and A.J. Parkin. "Improving the retention of aurally presented information." In *Practical Aspects of Memory*, Vol. 1, edited by M.M. Gruneberg, P.E. Morris, and R.N. Sykes, pp. 490–493, Wiley, 1998.

[8] J.C. Almenara. *La Formación en Internet: Guía para el Diseño de Materiales Didácticos.* Alcalá de Guadaira (Sevilla), España: Mad Ediciones, 2005.

[9] C. Andradas and E. Zuazua. Informe sobre la investigación matemática en España en el período 1990-1999. Technical report, CEAMM, 2002. Available at [url:160].

[10] E. André and T. Rist. "Controlling the behavior of animated presentation agents in the interface: Scripting versus instructing." *AI Magazine*, 22:4 (2001), 53–66.

[11] Argonne national labs access grid site. Available at [url:63].

[12] W.Y. Arms, C. Blanchi, and E.A. Overly. An architecture for information in digital libraries. *D-Lib Magazine* (1997). Available at [url:95].

[13] Art, Complexity and Technology: Their Interaction in Emergence. Workshop held in Villa Gualino, ISI Foundation, Torino, 5–6 May, 2005. Available at [url:119].

[14] F. Arzarello, M. Francaviglia, and R. Servidio. "Gesture and body-tactile experience in the learning of mathematical concepts." In *Proc. 5th Int. Conf. APLIMAT*

2006 (Bratislava, 7–10 February), edited by M. Kovacova, pp. 253–259, Bratislava: Slovak University of Technology, 2006.

[15] N. Ashieh and C. Knoblock. "Wrapper generation for semi–structured internet source." *ACM SIGMOD Record* 26:4 (1997), 8–15.

[16] Visual Resources Association. VRA core version 3, 2002. Available at [url:180].

[17] R. Ausbrooks, S. Buswell, D. Carlisle, S. Dalmas, S. Devitt, A. Diaz, M. Froumentin, R. Hunter, P. Ion, M. Kohlhase, R. Miner, N. Poppelier, B. Smith, N. Soiffer, R. Sutor, and S. Watt. Mathematical Markup Language (MathML) Version 2.0 (Second Edition), W3C Recommendation 21 October 2003. World Wide Web Consortium (W3C), 2003. Available at [url:184].

[18] G.J. Badros, J.J. Tirtowidjojo, K. Marriott, B. Meyer, W. Portnoy, and A. Borning. "A constraint extension to scalable vector graphics." In *WWW '01: Proceedings of the 10th International Conference on World Wide Web*, pp. 489–498, New York: ACM Press, 2001.

[19] J. Ball and J. Borwein. "Access: Who gets what access, when and how?" MSRI Workshop, Berkeley, 2005. Available at [url:143].

[20] T. Banchoff. *Beyond the Third Dimension.* New York: W.H. Freeman & Co., Scientific American Library, 1990.

[21] T. Banchoff. "Interactive geometry and multivariable calculus on the internet." *Proceedings of the KAIST International Symposium on Enhancing Undergraduate Teaching of Mathematics (Daejeon, Korea)*, pp. 11–31, 2005. Reprinted by the American Mathematical Society in CBMS Issues in Mathematics Education, Vol. 14, pp. 17–32, 2007.

[22] T. Banchoff and D. Cervone. "An interactive gallery on the internet: Surfaces beyond the third dimension." *International Journal of Shape Modeling* 5:1 (1999), 7–22.

[23] T. Banchoff and D. Cervone. "A virtual reconstruction of a virtual exhibit." *Multimedia Tools for Communicating Mathematics*, edited by J. Borwein, M. Morales, K. Polthier, and J.F. Rodrigues, pp. 29–38, Springer-Verlag, 2002.

[24] M. Barr. "Where does the money go?" *Newsletter on Serials Pricing Issues* 229 (1999). Available at [url:126].

[25] J.M. Barrueco Cruz and I. Subirats Coll. "Open archives en las universidades españolas." In III REBIUN Workshop, Barcelona, 2003. Available at [url:18].

[26] M. Bartošek, M. Lhoták, J. Rákosník, P. Sojka, and O. Ulrych. DML-CZ: Czech Digital Mathematics Library, 2005. Available at [url:11].

[27] H. Bass, G. Bolaños, R. Seiler, M. Seppälä, and S. Xambó. "e-Learning Mathematics." In *Proceedings of the 2006 International Congress of Mathematicians (ICM2006, August 22-30, Madrid), Invited Lectures III*, pp. 1743–1768, Freiburg: European Mathematical Society Publishing House, 2006.

[28] J.E. Beall, A. Doppelt, and J.F. Hughes. "Developing an interactive illustration: Using java and the web to make it worthwhile." In *Computer Graphics - Proceedings of 3D and Multimedia on the Internet, WWW and Networks*, Bradford, UK, 1996.

[29] O. Benjelloun, A. Das Sarma, C. Hayworth, and J. Widom. "An introduction to ULDBs and the Trio system." *IEEE Data Engineering Bulletin, Special Issue on Probabilistic Databases* 29 (2006), 5–16.

[30] P. Bérard. "Documentation issues for mathematics in the digital age." In *68th IFLA Council and General Conference*, August 2002. Available at [url:116].

[31] T. Berners-Lee. Information management: A proposal (1989). Available at [url:182].

[32] T. Berners-Lee, J. Hendler, and O. Lassila. "The semantic web." *Scientific American Magazine*, May 2001.

[33] P.A. Bertacchini, E. Bilotta, M. Francaviglia, and P. Pantano, Eds. "Mathematics, art and cultural industry." In *Proceedings of the National Conference on Matematica, Arte e Industria Culturale (Cetraro, 19-21 maggio 2005)*. CD-Rom by M.G. Lorenzi, ESG, University of Calabria, Cosenza, 2005. Available at [url:23].

[34] Bibliotecas y Objetos Digitales. Available at [url:73].

[35] E. Bilotta, M. Francaviglia, and P. Pantano, Eds. "Applications of mathematics to cultural industry." In *Proceedings of the Minisymposium Applications of Mathematics to Cultural Industry, held in occasion of the Conference SIMAI 2004 (Venice, 23-24 September)*, 2004. CD-Rom by M.G. Lorenzi, AVR S.r.L., Cosenza, 2004.

[36] E. Bilotta and P. Pantano. "Matematica, musica e tecnologie: un trinomio possibile." In *Proceedings of Matematica senza Frontiere*, pp. 209–324. Quad. del Dip. di Matematica dell'Univ. di Lecce 2, 2003. Available at [url:22].

[37] J.S. Birman. "Scientific publishing: A mathematician's viewpoint." *Notices of the AMS* 47:7 (2000), 770–774.

[38] M. Bordons et al. La investigación matemática española de difusión internacional: estudio bibliométrico (1996–2001). Technical report, MEC-RSME, 2005. Available at [url:79].

[39] J. Borwein, M.H. Morales, K. Polthier, and J.F. Rodrigues, Eds. *Multimedia Tools for Communicating Mathematics*. Springer-Verlag, 2002.

[40] T. Bouche. "Introducing the mini-DML project." In *New Developments in Electronic Publishing, AMS/SMM Special Session, Houston, May 2004 & ECM4 Satellite Conference (Stockholm, June 2004)*, edited by H. Becker, K. Stange, and B. Wegner, pp. 19–29. FIZ Karlsruhe, 2005. Available at [url:109].

[41] T. Bouche, Y. Laurent, and C. Sabbah. "L'édition sans drame." *Gazette des mathématiciens* 108 (2006), 86–88.

[42] T. Bouche. "A pdfLaTeX-based automated journal production system." *TUGboat* 27:1 (2006), 45–50.

[43] W.L. Boyd and G.C. Vanderheiden. "The graphical user interface: Crisis, danger, and opportunity." *Journal of Visual Impairment and Blindness* 10 (1990), 498–502.

[44] T. Bray, J. Paoli, C.M. Sperberg-McQueen, E. Maler, and F. Yergeau. Extensible Markup Language (XML) 1.0 (Fourth Edition) W3C Recommendation 16 (August 2006), 2006. Available at [url:185].

[45] B. Buchberger, A. Crăciun, T. Jebelean, L. Kovács, T. Kutsia, K. Nakagawa, F. Piroi, N. Popov, J. Robu, M. Rosenkranz, and W. Windsteiger. "Theorema: Towards computer-aided mathematical theory exploration." *Journal of Applied Logic* 4:4 (2006), 470–504.

[46] P. Buneman, A. Chapman, and J. Cheney. "Provenance management in curated databases." In *Proc. of ACM SIGMOD Int'l Conf. on Management of Data (Chicago, Illinois)*, New York: ACM Press 2006.

[47] P. Buneman, S. Khanna, and W.C. Tan. "Why and where: A characterization of data provenance." In *Proc. of the Eighth Int'l Conf. on Database Theory (London, UK)*, edited by J. Van den Bussche and V. Vianu, LNCS 1973, pp. 316–330, Heidelberg: Springer, 2001.

[48] V. Bush. "As we may think." *Atlantic Monthly* 176:1 (1945), 101–108. Available at [url:172].

[49] G. Caglioti. *Simmetrie Infrante nella Scienza e nell'Arte*, Milan: CLUP, 1983.

[50] G. Cappellato and N. Sala. *Architettura della Complessità: la Geometria Frattale tra Arte, Architettura e Territorio*, Milan: Franco Angeli Editore, 2004.

[51] O. Caprotti, M. Seppälä, and S. Xambó. "Novel aspects of the use of ICT in mathematics education." In *Proceedings of the 2006 International Joint Conferences*

on Computer, Information, and Systems Science, and Engineering (CISSE 2006, December 4–14), 2006.

[52] O. Caprotti, M. Seppälä, and S. Xambó. "Using web technologies to teach mathematics." In *Proceedings of the 2006 International Conference of the Society for Information Technology and Teacher Education (SITE 2006, March 20/24, Orlando, FL)*, Vol. 2006, Number 1, pp. 2679–2684, 2006.

[53] D. Carlisle. "OpenMath, MathML, and XSL." *ACM SIGSAM Bulletin* 34 (2000), 6–11.

[54] D. Carlisle. "MathML on the web: Using XSLT to enable cross-platform support for XHTML and MathML in current browsers." In *International Conference MathML and Technologies for Math on the Web (Chicago, Illinois, USA, June 28-30)*, Hickory Ridge Conference Center, 2002.

[55] D. Carlisle, J. Davenport, M. Dewar, N. Hur, and W. Naylor. Conversion between MathML and OpenMath. The OpenMath Consortium, 2001.

[56] L. Catastini and F. Ghione. *Le Geometrie della Visione*. Milan: Springer, 2004.

[57] CEMAT. Available at [url:78].

[58] CervanTeX. Available at [url:21].

[59] CESGA. Available at [url:84].

[60] E. Cherhal-Cleverly and L. Heigeas. "Mathdoc and the electronic publishing of mathematics." In *ElPub Conference*, Leuven, Belgium: David Brown Book Co, 2005.

[61] L. Childers, T. Disz, R. Olson, M. Papka, R. Stevens, and T. Udeshi. "Access grid: Immersive group-to-group collaborative visualization." In *Proceedings of the 4th International Immersive Projection Technology Workshop*, Iowa State University, Ames, Iowa, 2000.

[62] S.-C. Chou, X.-S. Gao, and J.-Z. Zhang. "Automated production of traditional proofs for constructive geometry theorems." In *Proceedings of the Eighth Annual IEEE Symposium on Logic in Computer Science LICS*, edited by M. Vardi, pp. 48–56, Los Alamitos, CA: IEEE Computer Society Press, 1993.

[63] S.-C. Chou, X.-S. Gao, and J.-Z. Zhang. "Automated generation of readable proofs with geometric invariants, I. multiple and shortest proof generation." *Journal of Automated Reasoning* 17 (1996), 325–347.

[64] CINDOC. Available at [url:88].

[65] J. Clark. XSL Transformations (XSLT) Version 1.0, W3C Recommendation 16 November 1999. World Wide Web Consortium (W3C), 1999. [url:187].

[66] R.C. Clark and R.E. Mayer. "E-learning and the science of instruction." *Markwell Biochemistry and Molecular Biology Education* 31 (2003), 217–218.

[67] A. Cohen, H. Cuypers, E.R. Barreiro, and H. Sterk. "Interactive mathematical documents on the web." In *Algebra, Geometry and Software Systems*, edited by M. Joswig and N. Takayama, pp. 289–306, Springer Verlag, 2003.

[68] W. Cohen and W. Fan. "Learning pageindependent heuristics for extracting data from web pages." In *8th World Wide Web Conference*, pp. 1641–1652, 1999.

[69] Committee on Electronic Information and Communication of the International Mathematical Union. Some best practices for retrodigitization. Available at [url:83].

[70] Consolider Mathematica. Available at [url:137].

[71] K. Crispien and H. Petrie. "Providing access to GUI's for blind people using a multimedia system based on spatial audio presentation." In *Preprint from the 95th Convention of the Audio Engineering Society*. New York: Audio Engineering Society, 1993.

[72] CSIC. Available at [url:93].

[73] S. Dalmas, M. Gaëtano, and S. Watt. An OpenMath 1.0 implementation. In *Proceedings of the 1997 International Symposium on Symbolic and Algebraic Computation (ISSAC)*, pp. 241–248, 1997.
[74] DCMI—Dublin Core Metadata Initiative. Available at [url:13].
[75] DCMI Education Working Group. Available at [url:14].
[76] R. de la Llave. "La publicación electrónica de la investigación matemática: Apuntes para un debate sobre su economía y sociología." *La Gaceta de la RSME* 6:2 (2003), 307–350.
[77] Instituto Portugues de Museus, 2006. Available at [url:118].
[78] M. Dedò. *Forme: simmetria e topologia.* Bologna: Zanichelli, 1999.
[79] K. Dennis, G.O. Michler, G. Schneider, and M. Suzuki. "Automatic reference linking in distributed digital libraries." In *Conference on Computer Vision and Pattern Recognition Workshop, Madison, Wisconsin,* Vol. 3, pp. 1–26, 2003.
[80] M. Dewar. OpenMath: An overview. *ACM SIGSAM Bulletin* 34 (2000), 2–5.
[81] DIALNET. Available at [url:10].
[82] S. Diehl and J. Keller. "VRML with constraints." In *VRML '00: Proceedings of the fifth symposium on Virtual reality modeling language (Web3D-VRML),* pp. 81–86, New York: ACM Press, 2000.
[83] S. Dill, N. Eiron, D. Gibson, D. Gruhl, R. Guha, A. Jhingran, T. Kanungo, S. Rajagopalan, A. Tomkins, J. Tomlin, and J. Zien. "Semtag and seeker: Bootstrappping the semantic web via automated semantic annotation." In *12th World Wide Web Conference*, pp. 178–186, 2003.
[84] DIMACS. Satisfiability suggested format. Available at [url:1].
[85] DML project, final report, October 2004. Available at [url:124].
[86] DOAJ. Available at [url:99].
[87] J.Y. Douglas and A. Hargadon. "The pleasure principle: Immersion, engagement, flow." In *HYPERTEXT '00: Proceedings of the eleventh ACM Conference on Hypertext and Hypermedia,* pp. 153–160, New York: ACM Press, 2000.
[88] S. Draves. The Fnord manual (1991). Available at [url:12].
[89] R. Duval. "Geometry from a cognitive point of view." In *Perspectives on the Teaching of Geometry for the 21st Century,* edited by C. Mammana C. and V. Villani, pp. 37–52, Kluwer Academic Publishers, 1998.
[90] T. Ebihara and N. Mirenkov. "Self-explanatory software components for computation on pyramids." *Journal of Three Dimensional Images* 14:4 (2000), 158–163.
[91] T. Ebihara, N. Mirenkov, R. Nomoto, and M. Nemoto. "Filmification of methods and an example of its applications." *International Journal of Software Engineering and Knowledge Engineering* 15:1 (2005), 87–115.
[92] W.K. Edwards, E.D. Mynatt, and K. Stockton. "Access to graphical interfaces for blind users." *Interactions* 2 (1995), 54–67.
[93] R. Eixarch, D. Marquès, and S. Xambó. "WIRIS: An internet platform for the teaching and learning of mathematics in large educational communities." *Contributions to Science* 2:2 (2002), 269–276.
[94] A. Elduque, N. Kamiya, and S. Okubo. "(-1,-1)-balanced Freudenthal Kantor triple systems and noncommutative Jordan algebras." Available at [url:6] and [url:178]. *Journal of Algebra* 294 (2005), 19–40.
[95] M. Emmer, Ed. *Matematica e Cultura 2000–2005*, Milan: Springer-Verlag, 2000–2005.
[96] M. Emmer. *The Visual Mind: Art and Mathematics,* Cambridge, MA: The MIT Press, 1994.
[97] ERIC: Educational Resources Information Center. Available at [url:112].

[98] P. Ernest. "A model of the cognitive meaning of mathematical expressions." *British Journal of Educational Psychology* 57 (1987), 343–370.

[99] A. Neri et al. "All'Ombra del Vesuvio," "Speciali Divulgativi," Bologna: CINECA, 2006. Available at [url:159].

[100] B. Barras et. al. The Coq proof assistant. Reference manual, version 6.1, 1997. (Version 8.0, LogiCal Project, 2004.)

[101] J. Ewing. "Measuring journals." *Notices of the AMS* 53:9 (2006), 1049–1053. Available at [url:68].

[102] J. Ewing. "Twenty centuries of mathematics: Digitizing and disseminating the past mathematical literature." *Notices of the AMS* 49:7 (2002), 771–777.

[103] J. Ewing. "Predicting the future of scholarly publishing." In *Electronic Information and Communication in Mathematics*, edited by F. Bai and B. Wegner, LNCS 2730, Springer, 2003.

[104] J. Ewing. "Misdirection." In *Proceedings of the ECM 4 satellite conference (KTH, Stockholm, Sweden)*, 2004.

[105] J.J. Exposito. "The devil you don't know: The unexpected future of open access publishing." *First Monday* 9:8 (2004).

[106] University of California Southern Regional Library Facility. The History of Microfilm: 1839 to the Present. Available at [url:166].

[107] S. Fantoni. Master document of "Comunicare la scienza," "Speciali divulgativi." Bologna: CINECA, 2006. Available at [url:159].

[108] G. Faraco and M. Francaviglia. *La Matematica dell'Arte—Visione, Enumerazione, Misurazione, Rappresentazione ed Astrazione: dagli Oggetti alle Forme* (manuscript in preparation).

[109] G. Faraco and M. Francaviglia. "A course of mathematics in art." In *Proceedings of the Minisymposium "Applied Mathematics in Cultural Industry," at the SIMAI 2004 Conference Venice, Italy, (23-24 September, 2004)*, edited by E. Bilotta, M. Francaviglia, and P. Pantano, AVR S.r.L., Cosenza, 2004.

[110] G. Faraco and M. Francaviglia. "Mathematical structures in art: A pathway for cultural industry." In *Creativity and Sciences, Vols. 4 and 5. Proceedings of 4th Understanding and Creating Music 2004 Conference (Caserta, November 23-26, 2004)*, edited by E. Bilotta et al., pp. 65–69, Caserta: Seconda Università di Napoli, 2005 (CD-ROM).

[111] G. Faraco and M. Francaviglia. "La geometria: una chiave di lettura per l'arte—prospettiva e simmetria." In *Proceedings of the National Conference "Matematica, Arte e Industria Culturale" (Cetraro, 19-21 May, 2005)*, edited by P.A. Bertacchini, E. Bilotta, M. Francaviglia, and P. Pantano, pp. 1–12, 2005. CD-ROM by M.G. Lorenzi, ESG, Univ. della Calabria, Cosenza, 2005.

[112] G. Faraco and M. Francaviglia. "Using art for mathematics teaching." In *Proceedings 4th Int. Conf. APLIMAT 2005 (Bratislava, February 1-4, 2005)*, edited by M. Kovacova, pp. 134–152, Bratislava: Slovak University of Technology, 2005.

[113] G. Faraco and M. Francaviglia. "From Menone to Heisenberg: A pedagogical path for non-specialist public." In *Proceedings 5th International Conference APLIMAT 2006 (Bratislava, 7-10 February 2006)*, edited by M. Kovacova, pp. 329–334, Bratislava: Slovak University of Technology, 2006.

[114] G. Farin. *Curves and Surfaces for CAGD—A Practical Guide.* Fifth edition. San Diego: Academic Press, 2002.

[115] L. Fatibene, M. Francaviglia, and M.G. Lorenzi. "Faster than light: Visualizing relativistic spacetime," a booklet accompanying a CD-ROM with video and multimedia called $E{=}mc^2$, aimed at divulgating the Theory of Special Relativity. University of Calabria Press (to appear).

[116] L. Fatibene, M. Francaviglia, and M.G. Lorenzi. "E=mc^2: A video and multimedia to visualize relativity." In *Proceedings of the 42nd Karpacz School on Theoretical Physics (Ladek Zdròj, February 2006)*, edited by G. Allemandi, A. Borowiec, and M. Francaviglia, Special Volume of *Intern. Journal of Geometric Methods in Modern Physics*, Singapore, 2007.

[117] L. Fatibene, M. Francaviglia, M.G. Lorenzi, and S. Mercadante. "Più veloce della luce?" Poster presented at the Conference *L'Università: Ponte tra Scienza e Società* (organized by Agorà Scienze, University of Torino, 15-16 September), 2006. Available at [url:66].

[118] L. Fatibene, M. Francaviglia, M.G. Lorenzi, S. Mercadante, and P. Pantano. "Arte & Matematica: un nuovo percorso." Poster presented at the Conference *L'Università: Ponte tra Scienza e Società* (organized by Agorà Scienze, University of Torino, 15-16 September), 2006.

[119] FECYT. Available at [url:115].

[120] M. Fernandes, J.A. Martins, J.S. Pinto, P. Almeida, and H. Zagalo. DisQS—Web services based distributed query system. In *International Conference on Internet Technologies and Applications*, North Wales, UK, 2005.

[121] M. Francaviglia and M.G. Lorenzi. "Communicating science today." In *Proceedings of the Conference on Mathematical Physical Models and Engineering Sciences, Dedicated to P. Renno in the Occasion of his 70th Birthday*, edited by G. D'Anna et al., Napoli: Liguori Editore (to appear).

[122] M. Francaviglia, M.G. Lorenzi, and M. Petry. "The space between—Superstring installation III." In *Proceedings of 8th Generative Art Conference GA2005 (Milan, December 15-17 2005)*, edited by C. Soddu, pp. 265–276, Milan: Alea Design Publisher, 2005.

[123] L.N. Gasaway. "Values conflict in the digital environment: Librarians versus copyright holders, 2000." Article based upon a speech given by Professor Gasaway at the 2000 Horace S. Manges Lecture, delivered on March 7, 2000 at the Columbia University School of Law. Available at [url:175].

[124] K. Gherab and J.L. González Quirós. *El Templo del Saber: Hacia la Biblioteca Digital Universal*. Barcelona: Ediciones Deusto/Planeta DeAgostini, 2006.

[125] M. Goossens, S Rahtz, E.M. Gurari, R. Moore, and R.S. Sutor. *The LaTeX Web Companion: Integrating TeX, HTML, and XML*. Addison Wesley, 1999.

[126] J. Görke, F. Hanisch, and W. Straßer. "Live graphics gems as a way to raise repositories for computer graphics education." *SIGGRAPH 2005 Education Paper*, 2005.

[127] P. Greco. Il nuovo comunicatore. CINECA, Bologna, 2006. Available at [url:159].

[128] A. Greenberg and R. Colbert. Navigating the sea of research on video conferencing-based distance education. Available at [url:157], 2004.

[129] M. Grötschel. "Electronic publishing, intellectual property, and open access in mathematics: The position of the international mathematical union." Slides of a Berlin presentation available at [url:81], October 2003.

[130] Moving Picture Experts Group. MPEG-7 overview, 2004. Available at [url:85].

[131] D. Haank. "Is electronic publishing being used in the best interests of science? The publisher's view." In *Proceedings of the Second ICSU/UNESCO International Conference on Electronic Publishing in Science (Paris, 20–23 February 2001)*, pp. 127–130, 2001.

[132] I. Hargittai and M. Hargittai. "The universality of the symmetry concept." *Nexus—Architecture and Mathematics*, pp. 81–95, 1996.

[133] T. Hede and A. Hede. "Multimedia effects on learning: Design implications of an integrated model." In *Untangling the Web: Establishing Learning Links, Pro-*

ceedings ASET Conference, (Melbourne, July 7–10), edited by S. McNamara and E. Stacey, 2002.

[134] C. Hernandez. "Some remarks on the new schemes in electronic publishing in mathematics." In *New Developments in Electronic Publishing*, AMS/SMM Special Session, pp. 81–90, Houston, May 2004.

[135] T. Hirotomi and N. Mirenkov. "Multimedia representation of computation on trees." *Journal of Three Dimensional Images* 13:3 (1999), 146–151.

[136] J.D. Hobby. "Introduction to MetaPost." In *EuroTEX '92 Proceedings (Prague, Czechoslovakia)*, pp. 21–36, 1992.

[137] J.J. Hull, B. Erol, J. Graham, and D. Lee. "Visualizing multimedia content on paper documents: Components of key frame selection for video paper." In *Seventh International Conference on Document Analysis and Recognition ICDAR'03 (Edinburgh, Scotland)*, 2003.

[138] Hippocratic database active enforcement installation guide, version 1.0. Technical report, IBM, 2006. Available at [url:67].

[139] IEEE 1484 Learning Object Metadata. Available at [url:34].

[140] Open Archives Initiative. Open archives initiative. Protocol for metadata harvesting, v.2.0, 2004. Available at [url:151].

[141] A. Jackson. "The digital mathematics library." *Notices of the AMS* 50:8 (2003), 918–923.

[142] D. Jacqué. Access grid aids SARS patients. Available at [url:70].

[143] P. Jacquier-Roux. RUCHE, an editorial flow management tool, 2006. Available at [url:52].

[144] P. Janičić. "GCLC—A tool for constructive euclidean geometry and more than that." In *Proceedings of International Congress of Mathematical Software (ICMS 2006)*, edited by N. Takayama, A. Iglesias, and J. Gutierrez, LNAI 4151, Heidelberg: Springer-Verlag, 2006.

[145] P. Janičić and P. Quaresma. "System description: GCLCprover + GeoThms." In *IJCAR 2006*, edited by U. Furbach and N. Shankar, LNAI, pp. 145–150, Heidelberg: Springer-Verlag, 2006.

[146] W. Jochems, J.V. Merriënboer, and K. Koper, Eds. *Integrated E-learning. Implications for Pedagogy, Technology and Organization.* London: RoutlegeFalmer, 2004.

[147] M. Jost and B. Wegner. "Emis 2001—A world-wide cooperation for communicating mathematics online." In *Electronic Information and Communication in Mathematics*, edited by F. Bai and B. Wegner, LNCS 2730, Springer, 1999.

[148] B. Kahle, R. Prelinger, and M.E. Jackson. "Public access to digital material." *D-Lib Magazine* 7:10, (2001). Available at [url:96].

[149] R. Kahn and R. Wilensky. A framework for distributed digital objects services. 1995. Available at [url:90].

[150] K. Kaiser. Some recent issues on the business of journal publishing: An independent point of view. Available at [url:110].

[151] N. Kamiya. On quadratic equations and quadratic algebras. Japan Mathematical Education Society, 2006.

[152] K. Kanev and S. Kimura. "Digital information carrier." Patent Registration No. 3635374, Japan Patent Office.

[153] K. Kanev and S. Kimura. "Digital information carrier." International Patent Application Publication WO 2005/055129, World Intellectual Property Organization (WIPO), 2005.

[154] K. Kanev and S. Kimura. "Direct point-and-click functionality for printed materials." *The Journal of Three Dimensional Images* 20:2 (2006), 51–59.

[155] T. Kasai, H. Yamaguchi, K. Nagano, and R. Mizoguchi. "System description of the goal of it education based on ontology theory." *JIEICE D–I*, J88-D-I No.1:3–15, 2005 (in Japanese).

[156] E. Keliher. "My turn: Forget the fads—the old way works best." *Newsweek*, September 30, 2002.

[157] Key Curriculum Press. *The Geometer's Sketchpad—Dynamic Geometry Software for Exploring Mathematics.* Emeryville, CA: Key Curriculum Press, 2002.

[158] D. Kirshner. "The visual syntax of algebra." *Journal for Research into Mathematics Education* 20 (1989), 274–287.

[159] J. Knight. "Cornell axes Elsevier journals as prices rise." *Nature* 426:217 (2003). doi:10.1038/426217a.

[160] D.E. Knuth. "Mathematical typography." *Bull. Amer. Math. Soc. (N.S.)* 1:2 (1979), 337–372.

[161] D.E. Knuth. Letter to the editorial board of the Journal of Algorithms, 2003. Available at [url:60].

[162] M. Kraus. "Interactive graphics everywhere." *Mathematica in Education and Research* 8:2 (1999), 27–29.

[163] M. Kraus. Direct manipulation of parametrized graphics. Presentation at the 2001 Mathematica Developer Conference, Champaign, Illinois, 2001.

[164] M. Kraus. "Parametrized graphics in mathematica." In *Proceedings of the 2002 World Multiconference on Systemics, Cybernetics, and Informatics (SCI 2002, Orlando, Florida)*, Vol. XVI, pp. 163–168, 2002.

[165] M. Kraus. CAGD applets, 2005. Available at [url:179].

[166] Kramerius System. Digital Library of the Academy of Sciences of the Czech Republic. Available at [url:33].

[167] J.H. Larkin and H.A. Simon. "Why a diagram is (sometimes) worth ten thousand words." *Cognitive Science* 11 (1987), 65–99.

[168] S. Lawrence. "Free online availability substantially increases a paper's impact." *Nature - WebDebates*, 2005.

[169] K. LeFevre, R. Agrawal, V. Ercegovac, R. Ramakrishnan, Y. Xu, and D. DeWitt. "Limiting disclosure in hippocratic databases." In *Proc. of the 30th Int'l Conf. on Very Large Databases*, Toronto, Canada, 2004.

[170] Ex Libris. Ex Libris Aleph overview, 2006. Available at [url:114].

[171] L. Weeks. "Pat Schroeder's new chapter: The former congresswoman is battling for America's publishers." *Washington Post*, February 7, 2001, C01. Available at [url:189].

[172] W. Liu and R. Stevens. "The access grid as knowledge technology." In *Workshop on Advanced Collaborative Environments*, 2004. Available at [url:62].

[173] E. Macias-Virgós. "Un gran proyecto de cooperación internacional: la biblioteca digital de matemáticas." *La Gaceta de la RSME* 6:2 (2003), 351–366.

[174] E. Macias-Virgós. "Some digitization initiatives in Spain." In *New Developments in Electronic Publishing of Mathematics, (Stockholm, June 25–27, 2004)*, FIZ Karlsruhe, Eggenstein-Leopoldshafen, Germany, 2005.

[175] A. Maedche. *Ontology Learning for the Semantic Web.* Kluwer, Academic Publishers, 2002.

[176] R.E. Mayer and R. Moreno. *A Cognitive Theory of Multimedia Learning: Implications for Design Principles.* New York: Wiley, 1997.

[177] R.E. Mayer and R. Moreno. "A split-attention effect in multimedia learning: Evidence for dual processing systems in working memory." *Journal of Educational Psychology* 90 (1998), 312–320.

[178] R.E. Mayer and R. Moreno. "Nine ways to reduce cognitive load in multimedia learning." *Educational Psychologist* 38 (2003), 43–52.

[179] MCU, Ministerio de Cultura. Available at [url:139].

[180] J. Mears. Grid technology helps fight SARS. Available at [url:144].

[181] MEC. Plan Nacional de I+D+i. Available at [url:140].

[182] R. Melero. "Open access environment in Spain: How the 'movement' has evolved and current emerging initiatives." In *Workshop Open Access and Information Management, Oslo*, 2006. Available at [url:19].

[183] J.G. Van Merriënboer and P. Ayres. "Research on cognitive load theory and its design implications for e-learning." *Educational Technology, Research and Development* 53:3 (2005), 5–13.

[184] G.O. Michler. "Report on the retrodigitization project 'Archiv der Mathematik."' *Arch. Math.* 77 (2001), 116–128.

[185] G.O. Michler. "How to build a prototype for a distributed digital mathematics archive library." *Annals of Mathematics and Artificial Intelligence* 38 (2003), 137–164.

[186] R. Miner. "The importance of MathML to mathematics communication." *Notice of the AMS* 52:5 (2005), 532–538.

[187] N. Mirenkov, A. Vazhenin, R. Yoshioka, T. Ebihara, T. Hirotomi, and T. Mirenkova. "Active knowledge studio." In *Fifth International Conference on Information Systems Analysis and Synthesis ISAS'99*, pp. 349–356, Orlando, 1999.

[188] N. Mirenkov, A. Vazhenin, R. Yoshioka, T. Ebihara, T. Hirotomi, and T. Mirenkova. "Self-explanatory components: A new programming paradigm." *International Journal of Software Engineering and Knowledge Engineering* 11:1 (2001), World Scientific, 5–36.

[189] *MONET (Mathematics on the Net)*. Available at [url:42].

[190] W. Mora. "Gráficos 3D interactivos con Mathematica y LiveGraphics3D." *Revista Digital Matemática, Educación e Internet* 6:1 (2005). Retrieved March 18, 2008 from [url:86].

[191] R. Moreno and R.E. Mayer. "A learner-centered approach to multimedia explanations: Deriving instructional design principles from cognitive theory." *Interactive Multimedia Electronic Journal of Computer-Enhanced Learning* 2:2 (2000).

[192] J. Narboux. "A decision procedure for geometry in coq." In *Proceedings TPHOLS 2004*, Vol. 3223, Lecture Notes in Computer Science, Springer, Heidelberg, 2004.

[193] W.N. Naylor and S.M. Watt. "On the relationship between OpenMath and MathML." In *Electronic Proc. Internet Accessible Mathematical Communication (IAMC 2001)*. Available at [url:31].

[194] OASIS DocBook. Available at [url:100].

[195] OASIS. Web Services Business Process Execution Language (WSBPEL), 2006. Available at [url:148].

[196] C. Obrecht. Eukleides: A Euclidean Geometry Drawing Language. Available at [url:113].

[197] P. Odifreddi. "La geometria dell'arte." in *Il Convivio—giornale telematico di poesia, arte e cultura*, 2002. Available at [url:32].

[198] Mathematical Society of Japan. *Encyclopedic Dictionary of Mathematics*. Cambridge, MA: MIT Press, 1993.

[199] Committee on Electronic Information Communication of the International Mathematical Union. "Best current practices: Recommendations on electronic information communication." *Notices of the AMS* 49:8 (2002), 922–925.

[200] OpenDOAR. Available at [url:152].

[201] F. Pass, A. Renkl, and J. Sweller. "Cognitive load theory and instructional design: Recent developments." *Educational Psychologist* 38 (2003), 1–4.

[202] K. Pavlou and R.T. Snodgrass. "Forensic analysis of database tampering." In *Proc. of the 25th ACM SIGMOD Int'l Conf. on the Management of Data*, Toronto, Canada, 2002.

[203] K. Perlin. The web as a procedural sketchbook. Course Notes, SIGGRAPH 2005 Course 8, 2005.

[204] K. Polthier, S. Khadem, E. Preuß, and U. Reitebuch. "Publication of interactive visualizations with JavaView." In *Multimedia Tools for Communicating Mathematics*, edited by J. Borwein, M. Morales, K. Polthier, and J.F. Rodrigues, pp. 241–264, Springer-Verlag, 2002.

[205] Project Gutenberg. *History and Philosophy of Project Gutenberg*. Available at [url:48].

[206] P. Quaresma and P. Janicic. Framework for constructive geometry (based on the area method). Technical Report 2006/001, Centre for Informatics and Systems of the University of Coimbra, 2006.

[207] P. Quaresma and A. Pereira. "Visualização de construções geométricas." *Gazeta de Matemática* 151 (2006), 39–41.

[208] R. Ràfols and A. Colomer. *Diseño Audiovisual*. Barcelona: Editoral Gustavo Gili, 2003.

[209] S. Ranise and C. Tinelli. The SMT-LIB format: An initial proposal, 2003. Available at [url:28].

[210] A. Rapp. The digitization centre at Goettingen state and university library. In *International Seminar on Digitization: Experience and Technology*, Lisbon, Portugal, 2004.

[211] REVICIEN. Available at [url:158].

[212] Revistas científicas electrónicas: estado del arte. Technical report, CINDOC, 2004. Available at [url:171].

[213] A. Bultheel and J. Teugels. "To whom it may concern: SCI and Mathematics." Open letter, Brussels, October 2003. Available at [url:7].

[214] J. Richter-Gebert and U.H. Kortenkamp. *The Interactive Geometry Software Cinderella*. Berlin: Springer-Verlag, 1999.

[215] E.G. Rieffel. "The genre of mathematics writing and its implications for digital documents." In *32nd Hawaii International Conference on System Sciences (HICSS-32)*, edited by R. Sprague Jr., IEEE Computer Society, 1999.

[216] ROAR. Available at [url:4].

[217] E.M. Rocha and J.F. Rodrigues. "Mathematical communication in the digital era." *Bol. Soc. Port. Mat.* 53 (2005), 1–21.

[218] J.F. Rodrigues. "Revistas matemáticas portuguesas." *Bulletin of the SPM* 50 (2004).

[219] J. Rogness. "Interactive gallery of quadric surfaces." *Journal of Online Mathematics and Its Applications* 5 (2005).

[220] L. Rollet and P. Nabonnand. "Une bibliographie mathématique idéale ? Le Répertoire bibliographique des sciences mathématiques." *Gazette des mathématiciens* 92 (2002), 11–26.

[221] C. Rossi. "De la diffusion à la conservation des documents numériques." *Cahiers GUTenberg* 49 (2007), 47–61.

[222] R. Roxas and N. Mirenkov. "Cyber-film: A visual approach that facilitates program comprehension." *International Journal of Software Engineering and Knowledge Engineering* 15:6 (2005), World Scientific, 941–975.

[223] C. Russel. Libraries in Today's Digital Age: The Copyright Controversy. Available at [url:141].

[224] M. Saber and N. Mirenkov. "A visual representation of cellular automata-like systems." *Journal of Visual Languages and Computing* 15 (2004), Elsevier Science, 409–438.

[225] N. Sala. "Matamatica, arte e architettura." *Didattica delle Scienze e Informatica* 200 (1999).

[226] N. Sala and G. Cappellato. *Viaggio Matematico nell'Arte e nell'Architettura.* Milan: Franco Angeli Editore, 2003.

[227] J. Santos, C. Teixeira, and J.S. Pinto. "eABC: um repositorio institucional virtual." In *XATA 2005 : XML Aplicações e Tecnologias Associadas*, pp. 40–51, Universidade do Minho, 2005.

[228] "SATLIB: An online resource for research on SAT." In: *SAT 2000*, edited by I.P. Gent, H.v. Maaren, and T. Walsh, pp. 283–292, IOS Press, 2000. SATLIB available at [url:161].

[229] R.D. Schafer. *An Introduction to Nonassociative Algebras.* New York: Dover Publications Inc., 1995.

[230] D. Schattschneider. *Visioni della simmetria—i disegni periodici di M.C. Escher.* Bologna: Zanichelli, 1992.

[231] W. Scheonpflug. "The trade-off between internal and external information storage." *Journal of Memory and Language* 25 (1986), 657–675.

[232] W. Schweikhardt, C. Bernareggi, N.J. Baptiste, B. Encelle, and M. Gut. "Lambda: A european system to access mathematics with braille and audio synthesis." *Lecture Notes in Computer Science* 4061 (2006), 1223–1230.

[233] SCIELO. Available at [url:162].

[234] C.E. Shannon. "The mathematical theory of communication." *Bell System Technical Journal* 27 (1948), 379–423 and 623–656.

[235] C.E. Shannon and W. Weaver. *The Mathematical Theory of Communication.* Urbana, IL: University of Illinois Press, 1949.

[236] C.E. Sherrick, J.C. Craig, W. Schiff, and E. Foulke. *Tactual Perception: A Sourcebook.* Cambridge University Press, 1982.

[237] R.S. Smith. *Guidelines for authors of Learning Objects.* Austin, TX: The New Media Consortium, 2004. Available at [url:146].

[238] R. Snodgrass, S. Yao, and C. Collberg. "Tamper detection in audit logs." In *Proc. of the 30th Int'l Conf. on Very Large Databases*, 2004.

[239] L.A. Sobreviela. A reduce-based OpenMath ↔ MathML translator. *ACM SIGSAM Bulletin* 34 (2000), 31–32.

[240] C. Soddu. "New naturality: A generative approach to art and design." *Leonardo Magazine* 35 (2002), 93–97.

[241] P. Sojka. "DML-CZ: From Scanned Image to Knowledge Sharing." In *Proceedings of I-KNOW '05: 5th International Conference on Knowledge Management*, edited by K. Tochtermann and H. Maurer, pp. 664–672, Graz, Austria: Know-Center in cooperation with Graz University, Joanneum Research, and Springer, 2005.

[242] P. Sojka, R. Panák, and T. Mudrák. "Optical character recognition of mathematical texts in the DML-CZ project, 2006." Talk at CMDE2006, 15–18 August 2006, Aveiro, Portugal.

[243] M. Sorice. *L'industria culturale in italia.* Milan: Editori Riuniti, 1998.

[244] W. Sperber. "Automatic classification of mathematical papers," 2006. Talk at CMDE2006, 15–18 August 2006, Aveiro, Portugal.

[245] Society for Scholarly Publishing. Held's editorial on Sabo bill, 2003. Available at [url:167].

[246] R.D. Stevens. Principles for the design of auditory interfaces to present complex information to blind people. Technical report, Department of Computer Science, University of York, 1996.

[247] A. Storino, Web Editor and Project Manager. Ricerca Italiana. Website under the supervision of S. Rago, CINECA, Bologna, 2006. Available at [url:159].

[248] Study on the economic and technical evolution of the scientific publication markets in Europe. Technical report, European Commission. Directorate-General for Research, 2004. Available at [url:16].

[249] G. Sutcliffe. The TPTP problem library. Available at [url:94].

[250] I.E. Sutherland. "Sketchpad a man-machine graphical communication system." In *25 Years of DAC: Papers on Twenty-five Years of Electronic Design Automation*, pp. 507–524, ACM Press, 1988.

[251] M. Suzuki. "Refinement of digitized mathematical journals by re-recognition, 2006." Talk at CMDE2006, 15–18 August 2006, Aveiro, Portugal.

[252] M. Suzuki, F. Tamari, R. Fukuda, S. Uchida, and T. Kanahori. "INFTY—An integrated OCR system for mathematical documents." In *Proceedings of the 2003 ACM Symposium on Document Engineering (Grenoble, France)*, pp. 95–104, New York: ACM Press, 2003.

[253] Interactive Digital Library Resources Information System. History of digital libraries, 2003. Available at [url:91].

[254] Tecnociencia. Available at [url:170].

[255] L. Tedeschini-Lalli. "Locale/globale: Guardare Picasso con sguardo 'riemanniano."' In *Matematica e Cultura 2001*, edited by M. Emmer, pp. 223–239, Milan: Springer-Verlag, 2002.

[256] The Educator's Reference Desk: Resource Guides. Available at [url:103].

[257] The OpenMath Society. Available at [url:153].

[258] C. Thiele. "Knuth meets NTG members." *MAPS* 16 (1996), 38–49.

[259] B. Tognazzini. First principles of interaction design, 2003. Available at [url:71].

[260] S. Trimberger. "Combining graphics and a layout language in a single interactive system." In *DAC '81: Proceedings of the 18th Conference on Design Automation*, pp. 234–239, IEEE Press, 1981.

[261] Los archivos abiertos y los repositorios institucionales. Curso de verano, El Escorial, 3–7 julio, 2006. Available at [url:174].

[262] B.T. Vander Zanden, R. Halterman, B.A. Myers, R. McDaniel, R. Miller, P. Szekely, D.A. Giuse, and D. Kosbie. "Lessons learned about one-way, dataflow constraints in the garnet and amulet graphical toolkits." *ACM Transactions on Programming Languages and Systems* 23:6 (2001), 776–796.

[263] A. Vazhenin, N. Mirenkov, and D. Vazhenin. "Multimedia representation of matrix computations and data." *International Journal of Information Sciences* 141 (2002), 97–122.

[264] Website of ESG. Available at [url:24].

[265] Website of the Ph.D course "Psychology of programming and artificial intelligence." Available at [url:24].

[266] Website of the portal "Matematita." Available at [url:132].

[267] Website of the portal "MArs." Available at [url:35].

[268] B. Wegner. "Emani—A project for the long-term preservation of electronic publications in mathematics online." In *Electronic Information and Communication in Mathematics*, edited by F. Bai and B. Wegner, LNCS 2730, Springer, 2003.

[269] B. Wegner. "Emani—Leader and follower for the WDML." In *New Developments in Electronic Publishing*, AMS/SMM Special Session, pp. 161–169, Houston, May 2004.

[270] J. Weitzman, Ed. "Sabo bill sparks copyright controversy." *Open Access Now*, August 25, 2003. Available at [url:74].

[271] Westgrid access grid information. Available at [url:191].

[272] Wikipedia. Peer-to-peer, 2006. Available at [url:17].

[273] K. Williams. *Nexus—Architecture and Mathematics*. Firenze: Gli Studi 2, Dell'Erba, 1996.

[274] S. Wolfram. *The Mathematica Book*. Fifth edition, Champaign, IL: Wolfram Media, 2003.

[275] WYP 2005 website. Available at [url:155] and [url:194].

[276] S. Xambó. *Block Error-Correcting Codes: A Computational Primer*. Springer-Verlag, 2003. Available at [url:193].

[277] XT. Available at [url:121].

[278] Yale University Library. *Standard License Agreement*. Available at [url:125].

[279] Z. Ye and et. al. Geometry expert, 2004. Available at [url:59].

[280] R. Yoshioka and N. Mirenkov. "A multimedia system to render and edit self-explanatory components." *Journal of Internet Technology* 3:1 (2002), 1–10.

[281] R. Yoshioka and N. Mirenkov. "Visual computing within environment of self-explanatory components." *Soft Computing Journal* 7:1 (2002), 20–32.

List of URLs

[url:1] ftp://dimacs.rutgers.edu/pub/challenge/satisfiability/doc/satformat.tex
[url:2] http://alem3d.obidos.org/en/intro/
[url:3] http://alem3d.obidos.org/en/itinex/
[url:4] http://archives.eprints.org/
[url:5] http://arxiv.org/
[url:6] http://arxiv.org/abs/math/0404550/
[url:7] http://bms.ulb.ac.be/documents/scieng.pdf
[url:8] http://bnd.bn.pt/
[url:9] http://calculadora.edu365.com
[url:10] http://dialnet.unirioja.es/
[url:11] http://dml.muni.cz
[url:12] http://draves.org/fnord/fnord.html
[url:13] http://dublincore.org/
[url:14] http://dublincore.org/groups/education/
[url:15] http://ec.europa.eu/research/eurab/pdf/eurab_scipub_report_recomm_dec06_en.pdf
[url:16] http://ec.europa.eu/research/science-society/pdf/scientificpublication-study_en.pdf
[url:17] http://en.wikipedia.org/wiki/Peer-to-peer/
[url:18] http://eprints.rclis.org/archive/00000384/
[url:19] http://eprints.rclis.org/archive/00006668/
[url:20] http://europa.eu.int/information_society/activities/digital_libraries/doc/pt_comm_digital_libraries.pdf
[url:21] http://filemon.mecanica.upm.es/CervanTeX/
[url:22] http://galileo.cincom.unical.it/bilotta/bilotta.htm
[url:23] http://galileo.cincom.unical.it/convegni/WSArte/home.htm
[url:24] http://galileo.cincom.unical.it/esg/index.htm
[url:25] http://gallica.bnf.fr/
[url:26] http://gdz.sub.uni-goettingen.de/
[url:27] http://gentzen.mat.uc.pt/~EukleidesPT/
[url:28] http://goedel.cs.uiowa.edu/smt-lib/
[url:29] http://hilbert.mat.uc.pt/~geothms/

[url:30] http://hutchinson.belmont.ma.us/tth/
[url:31] http://icm.mcs.kent.edu/research/iamc01proceedings.html
[url:32] http://ilconvivio.interfree.it
[url:33] http://kramerius.lib.cas.cz/kramerius/Welcome.do
[url:34] http://ltsc.ieee.org/wg12/
[url:35] http://mars.unical.it/
[url:36] http://math-doc.ujf-grenoble.fr/RBSM/
[url:37] http://math-doc.ujf-grenoble.fr/RBSM/
[url:38] http://mathdoc.emath.fr/
[url:39] http://mathnet.preprints.org/
[url:40] http://matwbn.icm.edu.pl/
[url:41] http://minidml.mathdoc.fr/
[url:42] http://monet.nag.co.uk/cocoon/monet/index.html
[url:43] http://nsdl.org/
[url:44] http://oai.bn.pt/
[url:45] http://opac.porbase.org/
[url:46] http://portail.mathdoc.fr/GALLICA/
[url:47] http://print.google.com/
[url:48] http://promo.net/pg/history.html#thepgphil
[url:49] http://purl.pt/2104/
[url:50] http://purl.pt/404/
[url:51] http://purl.pt/index/pmath/
[url:52] http://ruchedemo.cedram.org/
[url:53] http://scholar.google.com/
[url:54] http://scholar.google.com/intl/en/scholar/publishers.html
[url:55] http://scholar.google.com/intl/en/scholar/libraries.html
[url:56] http://sinbad.ua.pt/
[url:57] http://wdml.org/
[url:58] http://wiris.upc.es/EVAM/
[url:59] http://woody.cs.wichita.edu/gex
[url:60] http://www-cs-faculty.stanford.edu/~knuth/joalet.pdf
[url:61] http://www-texdev.ics.mq.edu.au/l2h/docs/manual/
[url:62] http://www-unix.mcs.anl.gov/fl/flevents/wace/wace2004/papers/paper-liu.pdf
[url:63] http://www-unix.mcs.anl.gov/fl/research/accessgrid/
[url:64] http://www.aarms.math.ca/events/atlantic/
[url:65] http://www.accessgrid.org/
[url:66] http://www.agorascienza.unito.it/
[url:67] http://www.almaden.ibm.Com/software/projects/iis/hdb/Publications/papers/HDBEnforcementUserGuide.pdf
[url:68] http://www.ams.org/ewing/
[url:69] http://www.ams.org/mathscinet/
[url:70] http://www.anl.gov/Media_Center/logos21-2/grid.htm
[url:71] http://www.asktog.com/basics/firstPrinciples.html
[url:72] http://www.b-on.pt/
[url:73] http://www.bibliotecasdigitales.es/
[url:74] http://www.biomedcentral.com/openaccess/archive/?page=features&issue=3
[url:75] http://www.bs.dk/MarcXchange/
[url:76] http://www.cabri.com/
[url:77] http://www.cbc.ca/cp/Oddities/061109/K110902U.html
[url:78] http://www.ce-mat.org
[url:79] http://www.ce-mat.org/documentos/informe_csic.pdf

[url:80] http://www.cedram.org/
[url:81] http://www.ceic.math.ca/Information/031022MPGOpenAccess.ppt
[url:82] http://www.ceic.math.ca/News/topology-letter.pdf
[url:83] http://www.ceic.math.ca/Publications/retro_bestpractices.pdf
[url:84] http://www.cesga.es/content/view/740/1/lang,en/
[url:85] http://www.chiariglione.org/MPEG/standards/mpeg-7/mpeg-7.htm
[url:86] http://www.cidse.itcr.ac.cr/revistamate/contribuciones-v6-n1-set2005/ParametrizacionGraficos3D/index.html
[url:87] http://www.cinderella.de/
[url:88] http://www.cindoc.csic.es/eng/
[url:89] http://www.claymath.org/millennium/Poincare_Conjecture/perelman+expositions.php
[url:90] http://www.cnri.reston.va.us/home/cstr/arch/k-w.html
[url:91] http://www.coe.missouri.edu/~DL/iDLR/viewpaper.php?pid=21
[url:92] http://www.copyright.gov/legislation/dmca.pdf
[url:93] http://www.csic.es/index.do?lengua=en
[url:94] http://www.cs.miami.edu/~tptp/TPTP/TR/TPTPTR.shtml
[url:95] http://www.dlib.org/dlib/february97/cnri/02arms1.html
[url:96] http://www.dlib.org/dlib/october01/10contents.html
[url:97] http://www.dml.cz/
[url:98] http://www.doaj.org/
[url:99] http://www.doaj.org/doaj?func=subject&cpid=57
[url:100] http://www.docbook.org/oasis/
[url:101] http://www.dublincore.org/
[url:102] http://www.dublincore.org/documents/dces/
[url:103] http://www.eduref.org/
[url:104] http://www.eiro.eurofound.eu.int/2004/12/feature/eu0412205f.html
[url:105] http://www.emani.org/
[url:106] http://www.emis.de/ELibM.html
[url:107] http://www.emis.de/MATH/DI.html
[url:108] http://www.emis.de/MATH/JFM/
[url:109] http://www.emis.de/proceedings/Stockholm2004/bouche.pdf
[url:110] http://www.emis.de/proceedings/Stockholm2004/kaiser.pdf
[url:111] http://www.emis.de/ZMATH/
[url:112] http://www.eric.ed.gov/
[url:113] http://www.eukleides.org/
[url:114] http://www.exlibrisgroup.com/aleph.htm
[url:115] http://www.fecyt.es/
[url:116] http://www.ifla.org/IV/ifla68/papers/095-112e.pdf
[url:117] http://www.intbooks.org/
[url:118] http://www.ipmuseus.pt/
[url:119] http://www.isi.it/main.php?liv1=news&liv2=past
[url:120] http://www.isinet.com/
[url:121] http://www.jclark.com/xml/xt-old.html
[url:122] http://www.jstor.org/
[url:123] http://www.keypress.com/sketchpad/
[url:124] http://www.library.cornell.edu/dmlib/DMLreport_final.pdf
[url:125] http://www.library.yale.edu/~llicense/standlicagree.html
[url:126] http://www.lib.unc.edu/prices/1999/PRIC229.HTML
[url:127] http://www.loc.gov/ead/
[url:128] http://www.loc.gov/marc/

[url:129] http://www.loc.gov/stadards/mets/
[url:130] http://www.loc.gov/standards/sru/
[url:131] http://www.loc.gov/z3950/agency/
[url:132] http://www.matematita.it/
[url:133] http://www.matf.bg.ac.yu/~janicic/gclc/
[url:134] http://www.mathaware.org/mam/00/918/index.html
[url:135] http://www.math.brown.edu/~banchoff/art/PAC-9603/
[url:136] http://www.math.brown.edu/TFBCON2003/
[url:137] http://www.mathematica.unican.es/
[url:138] http://www.mathematik.uni-bielefeld.de/~rehmann/DML/dml_links.html
[url:139] http://www.mcu.es/roai/en/inicio/inicio.cmd
[url:140] http://www.mec.es/ciencia/jsp/plantilla.jsp?area=plan_idi&id=2
[url:141] http://www.michaellorenzen.com/eric/copyright.html
[url:142] http://www.minervaeurope.org/listgoodpract.htm
[url:143] http://www.msri.org/specials/dmlp/
[url:144] http://www.networkworld.com/news/2003/0526sargrid.html
[url:145] http://www.niso.org/
[url:146] http://www.nmc.org/
[url:147] http://www.numdam.org/
[url:148] http://www.oasis-open.org/
[url:149] http://www.oclc.org/worldcat/
[url:150] http://www.openarchives.org/
[url:151] http://www.openarchives.org/OAI/openarchivesprotocol.html
[url:152] http://www.opendoar.org/
[url:153] http://www.openmath.org/
[url:154] http://www.orcca.on.ca/MathML/
[url:155] http://www.physics2005.org/
[url:156] http://www.plos.org/downloads/oa_whitepaper.pdf
[url:157] http://www.polycom.com/
[url:158] http://www.revicien.net/
[url:159] http://www.ricercaitaliana.it/
[url:160] http://www.rsme.es/inicio/informem.pdf
[url:161] http://www.satlib.org/
[url:162] http://www.scielo.org/index.php?lang=en
[url:163] http://www.soros.org/openaccess/
[url:164] http://www.soros.org/openaccess/oajguides/html/business_converting.htm
[url:165] http://www.spm.pt/
[url:166] http://www.srlf.ucla.edu/exhibit/text/BriefHistory.htm
[url:167] http://www.sspnet.org/custom/news/details.cfm?id=143
[url:168] http://www.sub.uni-goettingen.de/gdz/
[url:169] http://www.teachers.ash.org.au/jeather/maths/dictionary.html
[url:170] http://www.tecnociencia.org/
[url:171] http://www.tecnociencia.es/e-revistas/especiales/revistas/pdf/e-revistas_informe.pdf
[url:172] http://www.theatlantic.com/doc/194507/bush/
[url:173] http://www.theeuropeanlibrary.org/
[url:174] http://www.ucm.es/BUCM/biblioteca/11807.php
[url:175] http://www.unc.edu/~unclng/Columbia-article3.htm
[url:176] http://www.unicode.org
[url:177] http://www.unimarc.net/
[url:178] http://www.unizar.es/galdeano/preprints/2004/preprint09.pdf

[url:179] http://wwwvis.uni-stuttgart.de/~kraus/LiveGraphics3D/cagd/
[url:180] http://www.vraweb.org/vracore3.htm
[url:181] http://www.w3.org/Graphics/SVG/
[url:182] http://www.w3.org/History/1989/proposal.html
[url:183] http://www.w3.org/Math/
[url:184] http://www.w3.org/TR/2003/REC-MathML2-20031021/
[url:185] http://www.w3.org/TR/2006/REC-xml-20060816/
[url:186] http://www.w3.org/TR/MathML2/
[url:187] http://www.w3.org/TR/xslt
[url:188] http://www.w3.org/XML/
[url:189] http://www.washingtonpost.com/ac2/wp-dyn/A36584-2001Feb7
[url:190] http://www.webalt.com/Calculus-2006/
[url:191] http://www.westgrid.ca/
[url:192] http://www.wiris.com
[url:193] http://www.wiris.com/cc/
[url:194] http://www.wyp2005.org/activities.html
[url:195] http://www.zentralblatt-math.org/
[url:196] http://www.zim.mpg.de/openaccess-berlin/berlindeclaration.html

Contributors

Almeida, Pedro (pma@ieeta.pt)—Dept. Electronics, Telecommunications and Informatics, University of Aveiro, Portugal.

Banchoff, Thomas (tfb@cs.brown.edu)—Dept. Math., Brown University, USA.

Bartošek, Miroslav (bartosek@ics.muni.cz)—Library and Information Centre, Institute of Computer Science, Masaryk University, Czech Republic.

Bernareggi, Cristian (cristian.bernareggi@unimi.it)—Universita' degli Studi di Milano, Italy.

Borbinha, José (jlb@ist.utl.pt)—INESC-ID—Instituto de Engenharia de Sistemas e Computadores, Lisboa, Portugal.

Borwein, Jonathan (jborwein@cs.dal.ca)—Faculty of Computing Science, Dalhousie University, Canada.

Bouche, Thierry (thierry.bouche@ujf-grenoble.fr)—Institut Fourier & Cellule MathDoc, CNRS-UJF, Grenoble, France.

Breda, Ana (ambreda@mat.ua.pt)—Dept. Math., University of Aveiro, Portugal.

Brigatti, Valeria (info@valeriabrigatti.com)—Universita' degli Studi di Milano, Italy.

Campanozzi, Daniela (dany@dreamsangel.it)—Universita' degli Studi di Milano, Italy.

Caprotti, Olga (olga.caprotti@helsinki.fi)—University of Helsinki, Finland.

Ewing, John (jhe@ams.org)—American Mathematical Society, Providence, Rhode Island, USA.

Fernandes, Marco (marcopsf@ieeta.pt)—Dept. Electronics, Telecommunications and Informatics, University of Aveiro, Portugal.

Francaviglia, Mauro (mauro.francaviglia@unito.it)—Department of Mathematics, University of Torino, Italy.

Goto, Takatomo (gotoh@sm.u-tokai.ac.jp)—Dept. Math., Tokai University, Japan.

Janičić, Predrag (janicic@matf.bg.ac.yu)—Faculty of Mathematics, University of Belgrade, Serbia.

Jungic, Veselin (vjungic@sfu.ca)—Department of Mathematics, Simon Fraser University, Canada.

Kaiser, Klaus (kkaiser@uh.edu)—Houston Journal of Mathematics, Department of Mathematics, University of Houston, Texas, USA.

Kamiya, Noriaki (kamiya@u-aizu.ac.jp)—Center for Math. Sci., University of Aizu, Japan.

Kanev, Kamen (kanev@rie.shizuoka.ac.jp)—Research Institute of Electronics, Shizuoka University, Japan.

Kraus, Martin (krausma@in.tum.de)—Computer Graphics and Visualization, Technische Universität München, Germany.

Langstroth, David (dll@cs.dal.ca)—Faculty of Computing Science, Dalhousie University, Canada.

Lhoták, Martin (lhotak@lib.cas.cz)—Academy of Sciences Library, Czech Republic.

Lorenzi, Marcella G. (m.lorenzi@unical.it)—Evolutionary Systems Group, University of Calabria, Italy.

Macklem, Mason (mason@cs.dal.ca)—Faculty of Computing Science, Dalhousie University, Canada.

Macías-Virgós, Enrique (xtquique@usc.es)—Institute of Mathematics, Universidade de Santiago de Compostela, Spain.

Martins, Joaquim (jam@ieeta.pt)—Dept. Electronics, Telecommunications and Informatics, University of Aveiro, Portugal.

Messini, Andrea (andrea@dreamsangel.it)—Universita' degli Studi di Milano, Italy.

Mirenkov, Nikolay (nikmir@u-aizu.ac.jp)—Dept. Comp. Softw., University of Aizu, Japan.

Pantano, Pietro (PiePa@unical.it)—Department of Mathematics, University of Calabria, Italy.

Parreira, Telmo (tparreira@mat.ua.pt)—Geometrix Project, University of Aveiro, Portugal.

Pinto, Joaquim (jsp@ieeta.pt)—Dept. Electronics, Telecommunications and Informatics, University of Aveiro, Portugal.

Quaresma, Pedro (pedro@mat.uc.pt)—Dept. Math., University of Coimbra, Portugal.

Rákosník, Jiří (rakosnik@math.cas.cz)—Mathematical Institute, Academy of Sciences of the Czech Republic, Prague, Czech Republic.

Rocha, Eugénio M. (eugenio@ua.pt)—Dept. Math., University of Aveiro, Portugal.

Rodrigues, José F. (rodrigue@ptmat.fc.ul.pt)—University of Lisbon, Portugal.

Šárfy, Martin (sarfy@ics.muni.cz)—Library and Information Centre, Institute of Computer Science, Masaryk University, Czech Republic.

Seppälä, Mika (mika.seppala@helsinki.fi)—University of Helsinki, Finland & Florida State University, USA.

So, Clare (clso@maplesoft.com)—Maplesoft, Waterloo, Ontario, Canada.

Sojka, Petr (sojka@fi.muni.cz)—Dept. Computer Graphics and Design, Faculty of Informatics, Masaryk University, Czech Republic.

Sugita, Kimio (sugita@sm.u-tokai.ac.jp)—Dept. Math., Tokai University, Japan.

Tomašević, Jelena (jtomasevic@matf.bg.ac.yu)—Faculty of Mathematics, University of Belgrade, Serbia.

Tošić, Dušan (dtosic@matf.bg.ac.yu)—Faculty of Mathematics, University of Belgrade, Serbia.

Tsuchida, Kensei (kensei@toyonet.toyo.ac.jp)—Dept. Inf. and Comp. Sci., Tokyo University, Japan.

Vujošević-Janičić, Milena (milena@matf.bg.ac.yu)—Faculty of Mathematics, University of Belgrade, Serbia.

Watt, Stephen (watt@scl.csd.uwo.ca)—Ontario Research Centre for Computer Algebra, University of Western Ontario, London, Canada.

Wilson, Scott (swilson@cs.dal.ca)—Faculty of Computing Science, Dalhousie University, Canada.

Xambó, Sebastian (sebastia.xambo@upc.edu)—Technical University of Catalonia, Spain.

Yaku, Takeo (yaku@cs.chs.nihon-u.ac.jp)—Dept. Comp. Sci. and Sys. Analy., CHS, Nihon University, Japan.

Index